Phyto and Microbial Remediation of Heavy Metals and Radionuclides in the Environment

This book examines the role that bioremediation can play in the detoxification of soil, water, and air to improve environmental and human health, with a specific focus on heavy metals and radionuclides.

Environmental pollution, whether by natural or human causes, with industrial activities being a key player, is a challenge facing all nations across the world. While treatment has typically required the use of expensive technology, one promising solution is the use of phytoremediation, in which plants act by metabolizing or sequestering pollutants. This eco-friendly solution is a good alternative to the standard methods of soil and water treatments. This book provides not only the basic definitions and classification of technologies used for contaminant remediation but also the most recent studies dealing with the selection of new promising microbial stains and plant varieties involved in the treatment of radioactive and heavy metal contaminants. It provides a detailed description of the biochemical mechanisms and genes involved in the bioremediation of radionuclides and heavy metals, offering a clear insight for academics and practitioners interested in in vitro and *in situ* biological treatment.

This book will be of great value to students and scholars interested in environmental pollution and environmental health from across a range of different disciplines, including environmental microbiology and chemistry, ecology and environmental science, biological and environmental engineering and biotechnology.

Rym Salah-Tazdaït is Associate Professor (*Maître de Conférences classe A*) at the Department of Environmental Engineering, National Polytechnic School, Algiers, Algeria.

Djaber Tazdaït is Associate Professor (*Maître de Conférences classe A*) at the Department of Natural and Life Sciences, Faculty of Sciences, Algiers 1 University Benyoucef Benkhedda, Algiers, Algeria.

Routledge Focus on Environment and Sustainability

For more information about this series, please visit: www.routledge.com/ Routledge-Focus-on-Environment-and-Sustainability/book-series/RFES

Phyto and Microbial Remediation of Heavy Metals and Radionuclides in the Environment

An Eco-Friendly Solution for Detoxifying Soils

**Rym Salah-Tazdaït
and Djaber Tazdaït**

Routledge
Taylor & Francis Group

LONDON AND NEW YORK

from Routledge

First published 2022
by Routledge
4 Park Square, Milton Park, Abingdon, Oxon OX14 4RN

and by Routledge
605 Third Avenue, New York, NY 10158

Routledge is an imprint of the Taylor & Francis Group, an informa business

British Library Cataloguing-in-Publication Data
A catalogue record for this book is available from the British Library

Library of Congress Cataloging-in-Publication Data
A catalog record for this book has been requested

ISBN: 978-1-032-25305-3 (hbk)
ISBN: 978-1-032-25306-0 (pbk)
ISBN: 978-1-003-28260-0 (ebk)

DOI: 10.4324/9781003282600

Typeset in Times New Roman
by Apex CoVantage, LLC

This book is dedicated to our children, parents, and other family members.

Contents

Preface

The Industrial Revolution and the advent of new technologies have been historical turning points for an intensification of pollution, to the point that this constituted the beginning of the Anthropocene, the geological era defined by human impact on its environment. The concern for contaminated sites was born in late 1970 when they began to be considered toxic ticking time bombs. However, before the advent of a policy context requiring land decontamination, contaminated land often remained unused, mainly due to the prohibitive cost of decontamination.

The detoxification of soils and other environments (water and air) from potentially toxic compounds (metals, radioactive, or inorganic compounds) is challenging for all countries, independently of their development level. This critical purpose remains, however, difficult to achieve due to a lack of financial support.

The environment may become polluted with high concentrations of metals (cadmium, chromium, cobalt, copper, lead, mercury, nickel, selenium, zinc, etc.) and radioactive compounds (caesium, strontium, uranium, etc.) by a natural phenomenon or industrial activities. Its treatment by physical or chemical means often involves an expensive technology and requires site restoration.

Because the cost of conventional decontamination often exceeds the available budget or even the value of the land, and because engineering-based methods are not always aligned with the principles of development, sustainable, greener, and cheaper alternatives became necessary. One promising solution is the use of phytoremediation, in which plants act by metabolizing or sequestering pollutants. Three processes are observed: phytotransformation, phytostimulation (involving the stimulation of microbial biodegradation in the rhizosphere), and phytostabilization. This eco-friendly solution offers a good alternative to the common methods of soil and water treatments.

In this respect, this book presents a state-of-the-art analysis of the biological approaches applied to the remediation of heavy metals and radionuclides in polluted waters and soils. It explores the concept of bioremediation; bioremediation is the use of vascular plants, algae (phytoremediation), bacteria, or fungi (mycoremediation) to soil remediation, wastewater treatment, or indoor air sanitation, using plants and microorganisms. In particular, we will define the functions of plants and microorganisms in ecosystems, the various types of environmental contamination, and the various mechanisms of action of bioremediation: phytoremediation (phytoextraction, phytostabilization, phytodegradation, plant volatilization, or rhizofiltration) and microbial remediation (biotransformation, biosorption). We will also be interested in various applied projects worldwide, describing some successful *in situ* bioremediation applications for heavy metals and radionuclides immobilization. Finally, a brief focus will be dedicated to the integration in a circular economy of the bioremediation approach.

This book is aimed at both students and professionals; it is intended to be a definitive reference for students, teachers, agronomists, biologists, ecologists, foresters, and environmental practitioners.

Acknowledgements

The authors thank the publisher for the guidance and support in the compilation of this book.

Figures and tables

Figures

Tables

Abbreviations

APC	adenomatous polyposis coli gene
ArsR	arsenic transcriptional repressor gene
asd	aspartate-beta-semi-aldehyde dehydrogenase gene
Bax	BCL-2-associated-x protein
Bcl-2	B-cell lymphoma 2 protein
*BRCA*1/2	breast cancer gene 1/2
colE3	colicin E3 gene
DNA	deoxyribonucleic acid
DPC4	deleted in pancreatic carcinoma, locus 4
E_{h7}	oxidation-reduction potential at ph 7
$E°$	reduction potential under standard conditions
Ga	gallium
gef	guanine nucleotide exchange factor gene
GlpF	glycerol uptake facilitator protein gene
immE3	immunity E3 protein gene
In	indium
lacI	repressor of the lactose operon
LMP-*1*	latent membrane protein 1
Lu	lutetium
merA	mercuric reductase gene
NF2	neurofibromatosis 2
rat	rat glucocorticoid receptor gene
RB1	retinoblastoma gene 1
rolC	cytokinin-beta-glucosidase gene
URG4/URGCP	UP-regulated gene 4/upregulator of cell proliferation
xylS	alpha-xylosidase gene
Y	yttrium

Introduction

Introduction

With the acceleration of economic development, the human being is increasingly responsible for environmental pollution. In addition, the diversity of products of industrial origin leads to a considerable increase in the number of substances foreign to the living world, called xenobiotics. Some of these pollutants have contaminated soils and come from landfills, industrial facilities due to transport accidents, urban and industrial discharges, or agricultural practices.

Heavy metals (cadmium (Cd), copper (Cu), mercury (Hg), lead (Pb), zinc (Zn), etc.), metalloids (selenium (Se), arsenic (As), etc.), and toxic radionuclides persist in the environment and inevitably accumulate. They can migrate to surface water or groundwater or enter the food chain via plants to end up in animals and possibly in humans. When the risk to ecosystems and humans is very high, depollution is required.

Depollution techniques can be subdivided into three broad categories: containment, physicochemical methods of decomposition of pollutants, and bioremediation. However, confining polluted sites with physical barriers does not eliminate toxic substances and requires monitoring and maintaining these barriers. The physicochemical methods of decomposition of pollutants are very effective for certain contaminants present in the environment but have disadvantages (interventions are cumbersome and expensive and generally leave sterile soil).

Bioremediation is the use of biological processes to remove industrial pollutants that contaminate the biogeochemical cycle of natural substances. It is an advantageous option to reduce the pressure on the environment. Bioremediation uses biological systems (microorganisms or plants) to reduce air, water, or soil pollution.

Microorganisms have been the only living things on our planet for nearly two billion years. They are present everywhere and in large numbers: in the value of a teaspoon of soil, there are many more bacteria than humans

DOI: 10.4324/9781003282600-1

on the Earth. Their diversity is immense: it is estimated that biodiversity on the Earth resides mainly with microorganisms. They are the invisible agents of essential functions that they alone can assure; these functions are so essential that the higher living beings (humans, animals, and plants) could not live and feed without their microbial procession. The degradation of many pollutants is done through the enzymatic equipment of certain microorganisms. Properties such as fertility, resilience, and purifying power attributed to the soil or the aquatic environment are properties related to their microflora. Therefore, microorganisms are necessary to ensure the balanced functioning, fertility, habitability, and sustainability of the environment in which we live.

Practically all plants establish fungal symbioses at the root level. Some plants, particularly legumins, establish root symbioses with nitrogen-fixing bacteria. There are hundreds of thousands of plant species on the surface of the globe that exhibit a great diversity of forms and lifestyles related to adaptive capacities that allow them to survive in specific biotopes.

To do this, the metabolic pathways that organisms use to grow or to obtain energy can be used for molecules of polluting substances. Complete biodegradation detoxifies pollutants up to the stage of carbon dioxide, water, and harmless mineral salts. Incomplete biodegradation may provide less toxic degradation products than the original pollutant, but not necessarily. Sometimes, a degradation product more toxic and more carcinogenic than the initial compounds can be obtained.

The phenomenon of biodegradation can appear spontaneously; the terms "intrinsic bioremediation" or "natural attenuation" are then used. However, very often, the natural conditions are not sufficiently favourable. This kind of situation can be improved by completing one or more of the necessary factors.

In the field of bioremediation, there will be an increasing tendency to observe the speed of natural biodegradation and to step in only if the natural activity is not enough to eliminate the pollutant quickly.

Known as phytoremediation, the use of plants for the extraction of toxic products from the soil (especially heavy metals and radionuclides) has moved from the conceptual phase to the commercial phase. Research has shown that some plants can accumulate heavy metals in their aerial parts with relatively high rates. The low costs associated with phytoremediation technology and the possibility of recycling certain metals explain the growing interest in its development.

In this regard, this book summarizes and analyses the state of the art within the physiological, biochemical, and microbiological aspects of the bioremediation approach applied to soils and waters polluted by wasted heavy metals (Cd, chromium (Cr(VI)), cobalt, Cu, Pb, Hg, nickel (Ni), Se,

Zn, etc.) and radionuclides (caesium, strontium, uranium, etc.), and deals with the integration in a circular economy of bioremediation.

This book will provide specific insights on the role of genetic and biotechnological strategies for enhancing bioremediation processes. Moreover, environmental factors limiting the bioremediation of some heavy metals and nuclides in the field will be reported and discussed.

1 General outline of bioremediation

Introduction

The need to clean up contaminated sites has led to the development of new environmental technologies to destroy xenobiotic compounds rather than accumulate them in landfills. Bioremediation is an option that offers the possibility of destroying or making pollutants less toxic by using natural biological activities. The technologies used for contaminant remediation include phytoremediation and microbial remediation (Tazdaït and Salah-Tazdaït 2021, 291).

Phytoremediation

Phytoremediation is the use of plants to eliminate or transform pollutants into less toxic compounds. Before remedying, the contamination must be assessed initially: determine the remediation objectives, including the contaminant level to be achieved, the type and use of soil to be remedied, the surface and depth of soil to be remedied, and the time and money allocated for remediation and biomonitoring. The most suitable plants should be tolerant of contamination and environmental conditions of the site and have sufficient biomass production compared to the objectives and a large accumulation in aerial parts or roots.

Although plants have long been used for soil remediation, important scientific discoveries during the last years have helped to improve the process and extend its scope (Davies, Cox, Robinson, and Pittman 2015, 1–5; Antoniadis, Levizou, Shaheen, Ok, et al. 2017, 621–45; Rizwan, Ali, Zia-ur-Rehman, Rinklebe, et al. 2018, 1175–86; Alvarez-Vázquez, Martínez, Rodríguez, Vázquez-Méndez, et al. 2019, 387–99; Agarwal, Sarkar, Chakraborty, and Banerjee 2019, 221–36; Jaskulak, Grobelak, and Vandenbulcke 2019; Al-Thani and Yasseen 2020, 1–12; Patra, Pradhan, and Patra 2020). It can be used against organic pollutants and inorganic pollutants in solid (soil), liquid (surface and groundwater), and gaseous media.

DOI: 10.4324/9781003282600-2

Phytoremediation includes the following processes (Figure 1.1):

1 Phytoextraction (phytoaccumulation) uses plants to extract bioavailable contaminants (especially heavy metals and radionuclides) from contaminated soils or waters. Phytoextraction of heavy metals and radionuclides occur by translocation, by sap, from the root to the shoot and leaf cells.
2 Rhizofiltration consists in adsorption or precipitation of pollutants at the plant roots.
3 Phytostabilization (phytosequestration) relies on plants which may minimize the mobility and availability of contaminants in the environment (retain contaminants in one place) to reduce the risk to human

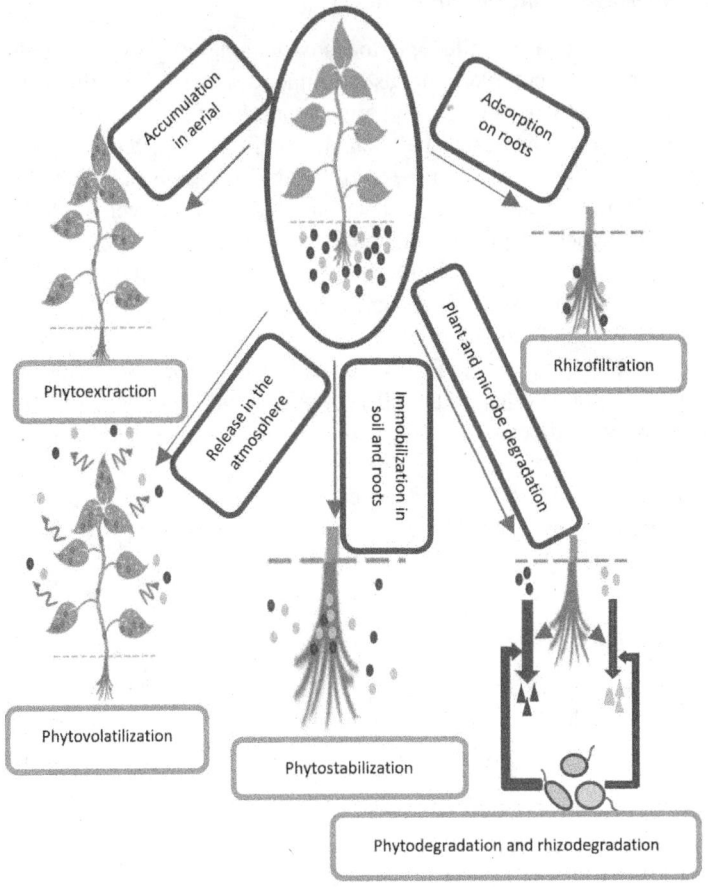

Figure 1.1 Phytoremediation techniques

health and the environment. Plants can absorb and neutralize contaminants by chelating them with biologically active particles.

4   Phytovolatilization (plant volatilization) is the uptake of a pollutant from the environment by roots and the release, by leaves, of a volatile form of an initially non-volatile pollutant or a volatile degradation product.

5   Phytodegradation (phytotransformation) and rhizodegradation are the breakdown of pollutants absorbed by plants through metabolic processes within the plant or the breakdown of pollutants by the microorganisms associated with the roots.

Phytoextraction (phytoaccumulation)

Phytoextraction, also called phytoaccumulation, uses hyperaccumulator plants (able to accumulate metals/radionuclides throughout their growth) to absorb pollutants from the soil and accumulate them in the aerial parts, which will then be harvested and incinerated (Table 1.1). Depending on their toxicity, the ashes would then be landfilled or better reused in metallurgy.

For example, the Chinese Ladder Brake *Pteris vittata* seems to be a very interesting hyperaccumulator plant due to its ability to accumulate simultaneously various elements, namely aluminium, "Al"; vanadium, "V"; nickel, "Ni"; cobalt, "Co"; selenium, "Se"; and uranium, "U" (Singh, Sudhakar, Swaminathan, and Rao 2014, 58).

Plants for phytoextraction must produce much above-ground biomass quickly; accumulate large quantities of metals/radionuclides in the above-ground parts; and be competitive and tolerant of contamination and severe living conditions.

This technique can use plants of economic interest, such as trees which will provide firewood. It is desirable to recycle plant biomass to recover the contaminants they contain.

Rhizofiltration

The rhizosphere is a dynamic region ruled by complex interactions between life forms in direct interdependence with roots. It is used for the elimination of pollutants from aquatic and terrestrial ecosystems by adsorption techniques using hydrophytes (plants living in water, such as Lemna (Duckweed), Pistia (Water Cabbage), Eichhornia (Water hyacinth), curly pondweed (*Potamogeton crispus*), Ceratophyllum, Hydrilla, and Eelgrass (*Vallisneria*)) and mesophytes (terrestrial plants adapted to neither particularly dry nor particularly wet environments, such as clover, corn (maize), goldenrod, lilac, oxeye daisy, and privet).

Table 1.1 Some examples of metal/radionuclide hyperaccumulator plants

Pollutants		Metal/radionuclide hyperaccumulator plants	Reference
Metals	Zn (zinc)	*Viola lutea ssp. calaminaria*	Sychta, Słomka, Suski, Fiedor, et al. 2018, 672
	Ni (nickel)	*Thlaspi oxyceras (Boiss.)*	Shahzad, Tanveer, Rehman, Cheema, et al. 2018
	Cu (copper)	*Astragalus patulum*	Antoniadis, Levizou, Shaheen, Ok, et al. 2017, 115
	As (arsenic)	*Pteris vittata*	Yan, Yiwei, Wu, Wang, et al. 2019, 386
	Cd (cadmium)	*Carthamus tinctorius*	Yadav, Gupta, Kumar, Reece, et al. 2018, 280
Radionuclides	^{230}Th (thorium)	*Juncus effusus*	Malhotra, Agarwal, and Gauba 2014, 77
	U (uranium)	*Callitriche stagnalis*	Pratas, Paulo, Favas, and Perumal 2014, 170
	^{226}Ra (radium)	*Medicago sativa* (alfalfa)	Alharbi and El-Taher 2013, 532
	^{210}Po (polonium)	*Glycine max* (soybean)	Galhardi, García-Tenorio, Bonotto, Francés, et al. 2017, 37
	^{238}Pu (plutonium)	*Solanum tuberosum* L. (potato plants)	Tawussi, Gupta, Mühr-Ebert, Schneider, et al. 2017, 186

Plants capable of efficient rhizofiltration must have rapidly growing roots and high biomass to obtain the greatest contact surface with the polluted matrix (Dushenkov, Kumar, Motto, and Raskin 1995, 1239). This is the case of the yellow flag flower *Iris pseudacorus* L. (Caldelas, Araus, Febrero, and Bort 2012, 1217), the common reed (*Phragmites australis*) (Weis and Weis 2004, 685; Ghassemzadeh, Yousefzadeh, and Arbab-Zavar 2008, 1668), the Indian mustard *Brassica juncea* (Moffat 1995, 302–3; Rizwan, Ali, Zia-ur-Rehman, Rinklebe, et al. 2018, 1176; Du, Guo, Li, Guo, et al. 2020), and sunflower (*Helianthus annuus* L.) (Dushenkov, Kumar, Motto, and Raskin 1995, 1239).

The fungi and bacteria that live in the rhizosphere use the exudates released by the roots for their growth and metabolic activity, which is why their number is two to four times greater in this region of the soil (Eisenhauer, Lanoue, Strecker, Scheu, et al. 2017, 3; Odelade and Babalola 2019, 4).

The mechanisms involved in the treatment of pollutants vary according to the type of pollutants and plants. We can quote biological processes or surface adsorption, physical and chemical processes such as chelation, ion exchange, and specific adsorption (Sharma, Singh, and Manchanda 2015, 955; Shahzad, Tanveer, Rehman, Cheema, et al. 2018).

Phytostabilization (phytosequestration)

By some processes, it is possible to immobilize metals/radionuclides in polluted soils and avoid contamination of runoff or groundwater, which could occur due to rainwater infiltration (Davies, Cox, Robinson, and Pittman 2015, 2). A high intake of organic compounds, iron hydroxides, or phosphate fertilizers such as phosphoric acid and calcium phosphate stimulates the formation of compounds that modify the chemical form and their bioavailability and thus contribute to the immobilization of metals/radionuclides (a form of inactivation of contaminants). Phytostabilization involves installing plants that tolerate toxic pollutants (metals/radionuclides) in the soil. They also limit erosion and prevent dust from entering the atmosphere. They can also secrete substances that chemically stabilize metals/radionuclides in the rhizosphere. Plants that can also accumulate metals/radionuclides in their root system are interesting for phytostabilization (Lack, Chaudhuri, Kelly, Kemner, et al. 2002, 2704; Gadepalle, Kesraoui-Ouki, Herwijnen, and Hutchings 2007, 233; Dave and Chopda 2014, 1; Kobets, Fedorova, Pshinko, Kosorukov, et al. 2014, 325; Martinez, Beazley, and Sobecky 2014, 1; von der Heyden and Roychoudhury 2015, 267–9; Yan, Zhou, and Liang 2015, 1–2; Mohamadiun, Dahrazma, Saghravani, and Khodadadi Darban 2018, 98–9).

Phytostabilization decreases the labile pool of contaminants by storing the contaminant in the root system or by promoting its insolubilization in the rhizosphere. It is not a pollution control in the strict sense because it changes the intensity and/or the chemical speciation in the exposure route without extracting the contaminant, but it allows a better landscape integration for sites with significant contamination and for which other methods are not applicable (Pilon-Smits 2005, 25; Rosatto, Roccotiello, Di Piazza, Cecchi, et al. 2019, 244). The effectiveness of this technique can be improved by making soil amendments that will immobilize the contaminants in the labile pool and fertilize the soil in some cases; this is called assisted phytostabilization (Mench, Vangronsveld, Beckx, and Ruttens 2006, 51; Lebrun, Miard, Nandillon, Légerc, et al. 2018, 318). This technique is broken down into two stages, the immobilization of contaminants and the establishment of tolerant plants. When adding amendments, consider how much to apply, how it affects other components, and how long term it is (Hettiarachchi, Pierzynski, and Ransom 2001, 1214–5). The best candidate plants for phytostabilization have a quick and easy establishment; have rapid growth; and are tolerant of contaminants and other factors such as low pH, salinity, coarse-textured soil, and drought. In addition, they must have significant plant cover to limit wind erosion and a low translocation of the metal to the upper parts, a long life cycle, a high reproduction rate, and commercial/economic advantages to make the contaminated site profitable (Mench, Vangronsveld, Beckx, and Ruttens 2006, 51). We must therefore choose the "excluders": plants that store preferentially in the roots. This technique is used when the contaminated soils have too high an intensity of exposure in the soil solution, and the phytoextraction would be very long. In the end, the mobility of the contaminant and its entry into the food chain are reduced, and the plant cover will accelerate the process of natural attenuation of the soil (Wong 2003, 777).

Phytovolatilization (plant volatilization)

In this case, the contaminants (metals/radionuclides) absorbed by plants are, during metabolism, associated with volatile compounds and released in the atmosphere (Pilon-Smits 2005, 19).

For example, the volatilization of Se from plant tissues can occur in the form of dimethyl diselenide $(CH_3)_2Se_2$ (Lewis, Johnson, and Broyer 1974, 107; Bañuelos, Zambrzuski, and Mackey 2000, 257) or in the form of dimethyl selenide $(CH_3)_2Se$ (Adler 1996, 43).

Another example concerns fluoride. Entry of inorganic fluorides into plants is possible via direct uptake through roots and airborne deposition via stomata. Fluoride can be accumulated in roots, leaves, and fruits (Bhargava and Bhardwaj 2011, 37; Weerasooriyagedara, Ashiq, Rajapaksha, Wanigathunge, et al. 2020). The calcium present in the cell wall can be a barrier to fluoride accumulation (Msagati, Mamba, Sivasankar, and Omine 2014, 236). Also, it has been observed that chloride deficiency accelerates the uptake of fluorides (Weerasooriyagedara, Ashiq, Rajapaksha, Wanigathunge, et al. 2020).

Nevertheless, phytovolatilization is considered as a negative mechanism in terms of phytoremediation since pollution is displaced from one compartment to another without quantitative reduction of the pollution load and sometimes without reducing the toxic power of these pollutants.

Phytodegradation (phytotransformation) and rhizodegradation

Phytodegradation (phytotransformation) and rhizodegradation correspond to the degradation of organic pollutants in the plant itself thanks to microbes associated with plants in their rhizosphere or their roots (mycorrhizae, bacterial endophytes). Plants can themselves degrade organic compounds through their enzymes. They catabolize them into inorganic compounds or degrade them into more stable forms that they can store. The most frequent enzymes are dehalogenases, oxygenases, peroxidases, peroxygenases, laccases, carboxylesterases, nitrilases, phosphatases, and nitroreductases. Degradation can take place in the aerial parts and roots, for example, for PolyChlorinated Biphenyl (PCB), Polycyclic Aromatic Hydrocarbons (PAHs), other hydrocarbons, herbicides, TriNitroToluene (TNT), TriChlorEthylene (TCE), and Methyl Tert-Butyl Ether (MTBE) (Pilon-Smits 2005, 16).

Rhizodegradation thus takes place thanks to microorganisms that colonize the rhizosphere (microbial growth area and exchange surface between soil, microorganisms, and plants). Microbial communities (bacteria, fungi, mycorrhizae) can act on the absorption and biodegradation of pollutants facilitating the subsequent absorption of primary and secondary pollutants by the plant (Tissut, Raveton, and Ravanel 2006, 489).

Microbial remediation

The need to clean up contaminated sites has led to the development of new environmental technologies to destroy xenobiotic compounds rather than accumulate them in landfills. Bioremediation is an option that offers the possibility of destroying or making pollutants less toxic by using natural biological activities. Microorganisms have been used for about a century for the treatment of wastewater and composts. The polluted sites are frequently contaminated with a mixture of very complex organic compounds such as mineral oils or industrial solvents. In addition, there are inorganic pollutants such as heavy metals and radionuclides.

The process of bioremediation by microorganisms consists of activating the natural ability that many microscopic organisms (bacteria, microalgae, fungi) possess to degrade pollutants into inert compounds, such as water and carbon dioxide. These organisms can be indigenous (already present in the polluted zone), or exogenous (added in the middle), or be taken from the contaminated site, cultivated in the laboratory, and reintroduced into the soil (bioaugmentation). Bioremediation generally takes place under aerobic conditions. However, the application of bioremediation systems under anaerobic conditions allows the degradation of several recalcitrant molecules. The following are the leading technologies used in bioremediation:

1   Bioaugmentation: introducing microbial cultures to the surface of the contaminated medium to increase the biodegradation of organic contaminants.

2   Biofiltration: using a biofilter to treat gaseous emissions (Ezeonu, Tagbo, Anike, Oje, et al. 2012, 9).

3   Biostimulation: by stimulating the activity of native microbial populations (present in soil or groundwater) by providing nutrients and adjusting environmental conditions (oxidation-reduction potential, humidity) (Kumar and Gopal 2015, 871).

4   Composting: by fermentation of organic household garbage (food waste) and green waste (foliage, garden waste) to produce reusable compost in agriculture or in the garden to fertilize the land (Kumar and Gopal 2015, 869).

5   Bioleaching: solubilizing heavy metals using acidophilic bacteria operating in the presence or the absence of oxygen (Salinas-Martinez, de los Santos-Cordova, Soto-Cruz, Delgado, et al. 2008, 115; Vijayanand, Prabu, Raj, and Achary 2012, 55).

Technologies of microbial remediation

Technologies of microbial remediation are bioaugmentation, biofiltration, biostimulation, composting, and bioleaching.

Bioaugmentation

This technology consists of introducing cultures of microorganisms on the surface of the contaminated medium to increase the biodegradation of organic contaminants. Generally, the microorganisms are selected based on their ability to degrade the organic compounds present in the site to be depolluted. The culture may include one or more species of microorganisms (Figure 1.2). Nutrients are generally provided in the solution

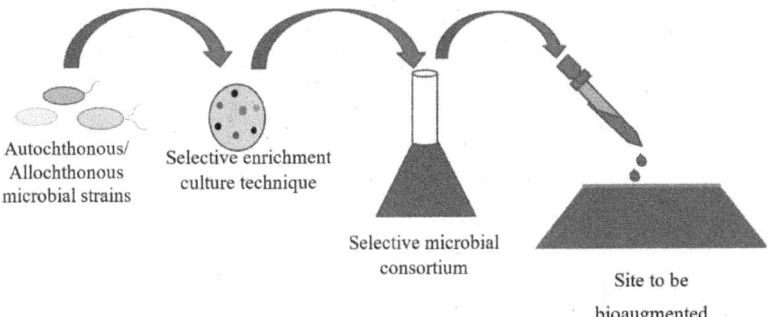

Autochthonous/
Allochthonous
microbial strains

Selective enrichment
culture technique

Selective microbial
consortium

Site to be
bioaugmented

Figure 1.2 Bioaugmentation technique

containing the microorganisms. This microorganism suspension is brought to the soil surface under natural conditions or injected into the contaminated site under pressure (Azubuike, Chikere, and Okpokwasili 2016, 12).

This technology is widely used to decontaminate sites containing hydrocarbons: the microorganisms chosen are bacteria with a great capacity for digestion of these hydrocarbons (Wu, Dick, Li, Wang, et al. 2016, 158). Additionally, efforts have been undertaken successfully to apply this technology to metal and radionuclide bioremediation. In a study by Wen, Wang, Li, Chen, et al. (2018, 481), three microbial isolates (*Enterobacteria* sp., *Rhodotorula mucilaginosa*, and *Meyerozyma guilliermondii*) were successfully tested as bioaugmentation agents for the treatment of an electroplating wastewater containing Cu^{2+}. In another study, the inoculation of the fungus *Aspergillus niger* improved the potential of *Syngonium podophyllum* (Arrowhead plant) to remove uranium from a simulated wastewater (Chao, Yin-hua, De-xin, Guang-yue, et al. 2019, 318).

Biofiltration

Biofiltration involves using a biofilter to treat gas emissions. The principle entails the use of microorganisms to degrade the pollutants in the air to be treated: the contaminated air is brought into contact with an aqueous phase in which the microbial population, also known as biomass, develops. In a biofiltration unit, the air to be purified (to clean up) first passes through a filter and a humidifier to remove the particles (dust, grease) present in the gas and bring the humidity level to 100%. The air is then introduced into a reactor (a tank) containing a lining formed of very porous materials (very greedy for humidity).

On the surface of the particles which constitute the lining is a biofilm that corresponds to a film of water-containing microorganisms (bacteria and fungi) whose function is to degrade the pollutants present in the air to be treated (Figure 1.3).

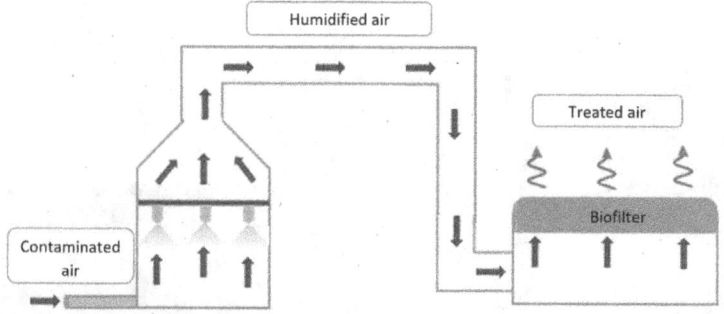

Figure 1.3 Biofiltration technique

This technology is, for example, used to treat air polluted by xylene or by nitrogenous compounds (Pujol and Tarallo 2000, 65; Wang, Nie, Luo, Yang, et al. 2015, 3815; Singh, Giri, Sahi, Geed, et al. 2017, 351). It was also applied to the removal of different heavy metals. For instance, it was possible to remove mercury (Hg°) through biosorption and possible bio-transformation occurring in biofilters containing different mixed bacterial cultures (autotrophic-denitrifying bacteria, sulphur-oxidizing bacteria, and toluene-degrading bacteria) (Philip and Deshusses 2008, 412).

Biostimulation

Biostimulation involves stimulating the activity of indigenous microbial populations (present in the soil or groundwater) by supplying nutrients and by adjusting the conditions of the environment (redox potential, humidity) (Coulon and Delille 2003, 469; Louati, Ben Said, Soltani, Got 2014, 3670; Ijoma, Selvarajan, Oyourou, Sibanda, et al. 2019, 542). To date, several research studies have dealt with the biostimulation of heavy metals and radionuclides-contaminated media with organic amendment. Thus, the effectiveness of using acetate for uranium removal by sulphate-reducing bacteria from sediment contaminated with uranium, selenium, vanadium, and so on has been reported by Xu, Veeramani, Qafoku, Singh, et al. (2017, 175).

Composting

Composting can be defined as a controlled biological process that ensures the transformation and recovery of organic matter (by-products of biomass, organic waste of biological origin) into a stabilized, hygienic product. It involves mixing excavated soil with organic soil improvers (called compost) and placing them in regularly spaced trapezoidal piles (swaths) to promote biodegradation. Aeration and humidity are two essential elements to maintain the conditions for good fermentation (Kästner and Miltner 2016, 3; Fan, Klemeš, Lee, and Ho 2018, 6; Ren, Zeng, Tang, Wang, et al. 2018, 138).

Bioleaching

Bioleaching is the leaching favoured by the biological way (generally bacterial). It corresponds to a methodology for the solubilization of heavy metals using acidophilic bacteria operating in the presence or absence of oxygen. Two factors are important for bioleaching: the temperature, which must be between 25 and 35°C, and the size of the particles, which must be very close to that of bacteria (Zhu, Zhang, Li, Han, et al. 2014, 2969; Borja, Nguyen, Silva, Park, et al. 2016, 128; Ghosh and Das 2017, 77; Wu, Liu, Zhang,

Zhu, et al. 2018, 1). This approach has been recently applied in printed circuit board treatment using the bacterium *Streptomyces albidoflavus* TN10 isolated from termite nest. The tested isolate showed a high potential in recovering different heavy metals, including Al, Zn, Ag, Ni, Fe, Cd, Cu, Pb, Ca (Kaliyaraj, Rajendran, Angamuthu, Antony, et al. 2019, 6).

Microorganisms used in bioremediation

They come from a wide variety of environments and can live in extreme conditions: temperatures below 0°C or, on the contrary, very high, in flooded environments or the middle of the desert, in the presence of an excess of oxygen or in anaerobic environment. Due to their adaptive power, these microorganisms are used to eliminate xenobiotic compounds. Among the aerobic bacteria known for their degrading power, we can cite those belonging to the genera *Pseudomonas*, *Alcaligenes*, *Sphingomonas*, and *Mycobacterium*. They can degrade pesticides, hydrocarbons, alkanes, and polyaromatic compounds. Often, they use the pollutant as a source of carbon and energy. Anaerobic bacteria are less common than aerobes. However, they are of great interest in the bioremediation of polychlorinated polyphenyls, trichloroethylene, and 1,2 dichloroethane. In all cases, the operation involves checking the availability of the microorganisms and the permanent adjustment of the conditions of their effectiveness: quantity and type of nutrients, oxygen concentration, pH, temperature, and salinity.

Many pollutants and pesticides can be transformed or broken down by both physicochemical and biological reactions. However, decomposition and transformation are generally faster in environments with microflora than in sterile environments.

In general, bacteria, actinomycetes, and fungi are involved in degrading pollutants in exposed soils. In submerged soils and anaerobic environments, bacteria are mainly involved. In the photic zone, cyanobacteria and microalgae can sometimes degrade certain pollutants, including pesticides, but are mainly involved in bioconcentration processes.

Many pollutants are present at very low concentrations in soils and waters. They are nonetheless dangerous, in particular, due to the phenomena of bioconcentration. The study of problems linked to the degradation of compounds present at very low concentrations is relatively recent and has highlighted the existence of a lower threshold below which the microflora can no longer multiply, and no degradation is observed. Conversely, too high a concentration of pollutants can cause inhibition of the microflora. The microbial specificity is, of course, correlated with the enzymatic specificity.

Before starting the degradation of a pollutant, there is generally a lag period that corresponds to an adaptation phase of the microflora. The duration of the latency phase can vary within wide limits between an hour and several months. It varies with the nature of the products but also with their concentration and environmental conditions. The repeated application of a pollutant quite often causes a shortening of the lag phase.

Also, many biodegradations, to be effective, require the combined action of several microorganisms (Kuhlmann and Schottler 1996, 289–94; Watanabe 2001, 237–9; Foulk and Bunn 2007, 438–43; Babaee, Bonakdarpour, Nasernejad, and Fallah 2010, 111–6; Das and Chandran 2011, 1–9; Jilani 2013, 257–63; Okere, Schuster, Ogbonnaya, Jones, et al. 2017, 1437–43; Nguyen, Chong, and Bui 2018, 2893).

Reference list

Adler, Tina. 1996. "Botanical cleanup crews". *Science News*, No. 3: 42–3. https://doi.org/10.2307/3980349.

Agarwal, Prerita, Mangaldeep Sarkar, Binayak Chakraborty, and Tirthankar Banerjee. 2019. "Chapter 7 – Phytoremediation of air pollutants: Prospects and challenges". In *Phytomanagement of Polluted Sites*, edited by Vimal Chandra, Pandey and Kuldeep Bauddh, 221–41. Cambridge, MA: Academic Press Elsevier Inc. https://doi.org/10.1016/B978-0-12-813912-7.00007-7.

Alharbi, Abdulaziz, and Atef M. El-Taher. 2013. "A Study on transfer factors of radionuclides from soil to plant". *Life Science Journal*, No. 2: 532–9. https://doi.org/10.7537/marslsj100213.78.

Al-Thani, R. F., and B. T. Yasseen. 2020. "Phytoremediation of polluted soils and waters by native Qatari plants: Future perspectives". *Environmental Pollution*, No. 259: 113694. https://doi.org/10.1016/j.envpol.2019.113694.

Alvarez-Vázquez, Lino J., Aurea Martínez, Carmen Rodríguez, Miguel E. Vázquez-Méndez, and Miguel A. Vilar. 2019. "Mathematical analysis and optimal control of heavy metals phytoremediation techniques". *Applied Mathematical Modelling*, No. 73: 387–400. https://doi.org/10.1016/j.apm.2019.04.011.

Antoniadis, Vasileios, Efi Levizou, Sabry M. Shaheen, Yong Sik Ok, Abin Sebastian, Christel Baum, Majeti N. V. Prasad, Walter W. Wenzel, and Jörg Rinklebe. 2017. "Trace elements in the soil-plant interface: Phytoavailability, translocation, and phytoremediation – A review". *Earth-Science Reviews*, No. 171: 621–45. https://doi.org/10.1016/j.earscirev.2017.06.005.

Azubuike, Christopher Chibueze, Chioma Blaise Chikere, and Gideon Chijioke Okpokwasili. 2016. "Bioremediation techniques – classification based on site of application: Principles, advantages, limitations and prospects". *World Journal of Microbiology and Biotechnology*, No. 11: 1–18. http://doi.org/10.1007/s11274-016-2137-x.

Babaee, R., B. Bonakdarpour, B. Nasernejad, and N. Fallah. 2010. "Kinetics of styrene biodegradation in synthetic wastewaters using an industrial activated sludge". *Journal of Hazardous Materials*, No. 1–3: 111–7. http://doi.org/10.1016/j.jhazmat.2010.08.012.

Bañuelos, G. S., S. Zambrzuski, and B. Mackey. 2000. "Phytoextraction of selenium from soils irrigated with selenium-laden effluent". *Plant and Soil*, No. 2: 251–8. https://doi.org/10.1023/A:1004881803469.

Bhargava, Devika, and Nagendra Bhardwaj. 2011. "Phytotoxicity of fluoride on a wheat variety (*Triticum aestivum* var. Raj. 4083) and its bioaccumulation at the reproductive phase". *Asian Journal of Experimental Sciences*, No. 25: 37–40.

Borja, Danilo, Kim Anh Nguyen, Rene A. Silva, Jay Hyun Park, Vishal Gupta, Yosep Han, Youngsoo Lee, and Hyunjung Kim. 2016. "Experiences and future challenges of bioleaching research in South Korea". *Minerals*, No. 4: 128. https://doi.org/10.3390/min6040128.

Caldelas, Cristina, Jose Luis Araus, A. Febrero, and Jordi Bort. 2012. "Accumulation and toxic effects of chromium and zinc in *Iris pseudacorus* L.". *Acta Physiologiae Plantarum*, No. 34: 1217–28. https://doi.org/10.1007/s11738-012-0956-4.

Chao, Zou, Sha Yin-hua, Ding De-xin, Li Guang-yue, Cui Yue-ting, Hu Nan, Zhang Hui, Dai Zhong-ran, Li Feng, Sun Jing, and Wang Yong-dong. 2019. "*Aspergillus niger* changes the chemical form of uranium to decrease its biotoxicity, restricts its movement in plant and increase the growth of *Syngonium podophyllum*". *Chemosphere*, No. 224: 316–23. https://doi.org/10.1016/j.chemosphere.2019.01.098.

Coulon, F., and D. Delille. 2003. "Effects of biostimulation on growth of indigenous bacteria in sub-antarctic soil contaminated with oil hydrocarbons". *Oil & Gas Science and Technology*, No. 4: 469–79. https://doi.org/10.2516/ogst:2003030.

Das, Nilanjana, and Preethy Chandran. 2011. "Microbial degradation of petroleum hydrocarbon contaminants: An overview". *Biotechnology Research International*, No. 941810. https://doi.org/10.4061/2011/941810.

Dave, Pragnesh N., and Lakhan V. Chopda. 2014. "Application of iron oxide nanomaterials for the removal of heavy metals". *Journal of Nanotechnology*, No. 398569: 1–14. http://doi.org/10.1155/2014/398569.

Davies, Helena S., Filipa Cox, Clare H. Robinson, and Jon K. Pittman. 2015. "Radioactivity and the environment: Technical approaches to understand the role of arbuscular mycorrhizal plants in radionuclide bioaccumulation". *Front Plant Science*, No. 580: 1–6. https://doi.org/10.3389/fpls.2015.00580.

Du, Juan, Zhanyu Guo, Ronghua Li, Di Guo, Altaf Hussain Lahori, Ping Wang, Xiangyu Liu, Xuejia Wang, Amjad Ali, and Zengqiang Zhang. 2020. "Screening of Chinese mustard (*Brassica juncea* L.) cultivars for Cd/Zn phytoremediation and research on physiological mechanisms". *Environmental Pollution*. https://doi.org/10.1016/j.envpol.2020.114213.

Dushenkov, Vyacheslav, Nanda P. B. A. Kumar, Harry Motto, and Ilya Raskin. 1995. "Rhyzofiltration: The use of plants to remove heavy metal from aqueous streams". *Environmental Science & Technology*: 1239–45. https://doi.org/10.1021/es00005a015.

Eisenhauer, Nico, Arnaud Lanoue, Tanja Strecker, Stefan Scheu, Katja Steinauer, Madhav P. Thakur, and Liesje Mommer. 2017. "Root biomass and exudates link plant diversity with soil bacterial and fungal biomass". *Scientific Reports*, No. 7: 44641 (April 2017). https://doi.org/10.1038/srep44641.

Ezeonu, Chukwuma S., Richard Tagbo, Ephraim N. Anike, Obinna A. Oje, and Ikechukwu N. E. Onwurah. 2012. "Biotechnological tools for environmental sustainability: Prospects and challenges for environments in Nigeria – A standard review". *Biotechnology Research International*, No. 450802: 1–26. https://doi.org/10.1155/2012/450802.

Fan, Yee Van, Jiří Jaromír Klemeš, Chew Tin Lee, and Chin Siong Ho. 2018. "Efficiency of microbial inoculation for a cleaner composting technology". *Clean Technologies and Environmental Policy*, No. 20: 517–27. https://doi.org/10.1007/s10098-017-1439-5.

Foulk, Jonn, and Joe M. Bunn. 2007. "Factors influencing the duration of lag phase during in vitro biodegradation of compression-molded, acetylated biodegradable soy protein films". *Journal of Food Engineering*, No. 79: 438–44. https://doi.org/10.1016/j.jfoodeng.2006.01.070.

Gadepalle, Vishnu Priya, Sabeha Kesraoui-Ouki, René Van Herwijnen, and Tony Hutchings. 2007. "Immobilization of heavy metals in soil using natural and waste materials for vegetation establishment on contaminated sites". *Soil and Sediment Contamination: An International Journal*, No. 2: 233–51. 10.1080/15320380601169441.

Galhardi, Juliana A., Rafael García-Tenorio, Daniel M. Bonotto, Inmaculada Díaz Francés, and João Gabriel Motta. 2017. "Natural radionuclides in plants, soils and sediments affected by U-rich coal mining activities in Brazil". *Journal of Environmental Radioactivity*, No. 177: 37–47. 10.1016/j.jenvrad.2017.06.001.

Ghassemzadeh, Fereshteh, Hadis Yousefzadeh, and M. H. Arbab-Zavar. 2008. "Removing arsenic and antimony by *Phragmites australis*: Rhizofiltration technology". *Journal of Applied Sciences*, No. 9: 1668–75. https://doi.org/10.3923/jas.2008.1668.1675.

Ghosh, S., and A. P. Das. 2017. "Bioleaching of manganese from mining waste residues using *Acinetobacter* sp.". *Geology, Ecology, and Landscapes*, No. 2: 77–83. https://doi.org/10.1080/24749508.2017.1332847.

Hettiarachchi, Ganga M., G. M. Pierzynski, and Michel D. Ransom. 2001. "In situ stabilization of soil lead using phosphorus". *Journal of Environmental Quality*, No. 30: 1214–21. 10.2134/jeq2001.3041214x.

Ijoma, Grace N., Ramganesh Selvarajan, Jean-Nazaire Oyourou, Timothy Sibanda, Tonderayi Matambo, Annie Monanga, and Kim Mkansi. 2019. "Exploring the application of biostimulation strategy for bacteria in the bioremediation of industrial effluent". *Annals of Microbiology Ann Microbiol*, No. 69: 541–51. https://doi.org/10.1007/s13213-019-1443-6.

Jaskulak, Marta, Anna Grobelak, and Franck Vandenbulcke. 2019. "Modelling assisted phytoremediation of soils contaminated with heavy metals – Main opportunities, limitations, decision making and future prospects". *Chemosphere*, No. 249: 126196. https://doi.org/10.1016/j.chemosphere.2020.126196.

Jilani, Seema. 2013. "Comparative assessment of growth and biodegradation potential of soil isolate in the presence of pesticides". *Saudi Journal of Biological Sciences*, No. 3: 257–64. https://doi.org/10.1016/j.sjbs.2013.02.007.

Kaliyaraj, Dhanalashmi, Shanmugasundaram Thangavel, Menaka Rajendran, Vignesh Angamuthu, Annam Renita Antony, Manigundan Kaari, Gopikrishnan Venugopal, Jerrine Joseph, and Radhakrishnan Manikkam. 2019. "Bioleaching of heavy metals from printed circuit board (PCB) by *Streptomyces albidoflavus* TN10 isolated from insect nest". *Bioresources and Bioprocessing*, No. 6: 1–11. https://doi.org/10.1186/s40643-019-0283-3.

Kästner, Matthias, and Anja Miltner. 2016. "Application of compost for effective bioremediation of organic contaminants and pollutants in soil". *Applied Microbiology and Biotechnology*, No. 100: 3433–49. https://doi.org/10.1007/s00253-016-7378-y.

Kobets, S. A., V. M. Fedorova, G. N. Pshinko, A. A. Kosorukov, and V. Ya. Demchenko. 2014. "Effect of humic acids and iron hydroxides deposited on the surface of clay minerals on the ^{137}Cs immobilization". *Radiochemistry*, No. 56: 325–31. https://doi.org/10.1134/S1066362214030175.

Kuhlmann, B., and U. Schottler. 1996. "Influence of different redox conditions on the biodegradation of the pesticide metabolites phenol and chlorophenols". *International Journal of Environmental Analytical Chemistry*, No. 1–4: 289–95. https://doi.org/10.1080/03067319608045562.

Kumar, Baduru Lakshman, and D. V. R. Sai Gopal. 2015. "Effective role of indigenous microorganisms for sustainable environment". *3 Biotech*, No. 6: 867–76. https://doi.org/10.1007/s13205-015-0293-6.

Lack, Joseph G., Swades K. Chaudhuri, Shelly D. Kelly, Kenneth M. Kemner, Susan M. O'Connor, and John D. Coates. 2002. "Immobilization of radionuclides and heavy metals through anaerobic bio-oxidation of Fe(II)". *Applied and Environmental Microbiology*, No. 6 (June 2002): 2704–10. 10.1128/AEM.68.6.2704-2710.2002.

Lebrun, Manhattan, Florie Miard, Romain Nandillon, Jean-Christophe Légerc, Nour Hattab-Hambli, Gabriella S. Scippa, Sylvain Bourgerie, and Domenico Morabito. 2018. "Assisted phytostabilization of a multicontaminated mine technosol using biochar amendment: Early stage evaluation of biochar feedstock and particle size effects on As and Pb accumulation of two *Salicaceae* species (*Salix viminalis* and *Populus euramericana*)". *Chemosphere*, No. 194: 316–26. https://doi.org/10.1016/j.chemosphere.2017.11.113.

Lewis, Barbara Ann Gamboa, C. M. Johnson, and T. C. Broyer. 1974. "Volatile selenium in higher plants the production of dimethyl selenide in cabbage leaves by enzymatic cleavage of Se-methyl selenomethionine selenonium salt". *Plant Soil*, No. 1: 107–18. https://doi.org/10.1007/BF00011413.

Louati, Hela, Olfa Ben Said, Amel Soltani, Patrice Got, Cristiana Cravo-Laureau, Robert Duran, Patricia Aissa, Olivier Pringault, and Ezzeddine Mahmoudi. 2014. "Biostimulation as an attractive technique to reduce phenanthrene toxicity for meiofauna and bacteria in lagoon sediment". *Environmental Science and Pollution Research International*, No. 5: 3670–9. https://doi.org/10.1007/s11356-013-2330-5.

Malhotra, Raveesha, Sahil Agarwal, and Pammi Gauba. 2014. "Phytoremediation of radioactive metals". *Journal of Civil Engineering and Environmental Technology*, No. 5: 75–9.

Martinez, Robert J., Melanie J. Beazley, and Patricia A. Sobecky. 2014. "Phosphate-mediated remediation of metals and radionuclides". *Advances in Ecology*, No. 786929: 1–14. http://doi.org/10.1155/2014/786929.

Mench, M., Jaco Vangronsveld, Carolien Beckx, and Ann Ruttens. 2006. "Progress in assisted natural remediation of an arsenic contaminated agricultural soil". *Environmental Pollution*, No. 1: 51–61. https://doi.org/10.1016/j.envpol.2006.01.011.

Moffat, Anne Simon. 1995. "Plants providing their worth in toxic metal cleanup". *Science*, No. 5222: 302–3. https://doi.org/10.1126/science.269.5222.302.

Mohamadiun, Malihe, Behnaz Dahrazma, Seyed Fazlolah Saghravani, and Ahmad Khodadadi Darban. 2018. "Removal of cadmium from contaminated soil using iron (III) oxide nanoparticles stabilized with polyacrylic acid". *Journal of Environmental Engineering and Landscape Management*, No. 2: 98–106. https://doi.org/10.3846/16486897.2017.1364645.

Msagati, Titus Alfred Makudali, Bhekie B. Mamba, Venkataraman Sivasankar, and Kiyoshi Omine. 2014. "Surface restructuring of lignite by bio-char of *Cuminum cyminum* – Exploring the prospects in defluoridation followed by fuel applications". *Applied Surface Science*, No. 301: 235–43. https://doi.org/10.1016/j.apsusc.2014.02.052.

Nguyen, Huong Lan, Min Nyuk Chong, and Ha Manh Bui. 2018. "Shortening the acclimation and degradation lag of xenobiotics by enriching the energy content of microbial populations". *Polish Journal of Environmental Studies*, No. 6: 2893–7. https://doi.org/10.15244/pjoes/81110.

Odelade, Kehinde Abraham, and Olubukola Oluranti Babalola. 2019. "Bacteria, fungi and archaea domains in rhizospheric soil and their effects in enhancing agricultural productivity". *International Journal of Environmental Research and Public Health*, No. 20: 1–50. https://doi.org/10.3390/ijerph16203873.

Okere, Uchechukwu V., Jasmin K. Schuster, Uchenna Ogbonnaya Ogbonnaya, Kevin C. Jones, and Kirk T. Semple. 2017. "Indigenous 14C-phenanthrene biodegradation in 'pristine' woodland and grassland soils from Norway and the United Kingdom". *Environmental Science: Processes Impacts*, No. 11: 1437–44. https://doi.org/10.1039/C7EM00242D.

Patra, Deepak Kumar, Chinmay Pradhan, and Hemanta Kumar Patra. 2020. "Toxic metal decontamination by phytoremediation approach: Concept, challenges, opportunities and future perspectives". *Environmental Technology & Innovation*, No. 18: 100672. https://doi.org/10.1016/j.eti.2020.100672.

Philip, Ligy, and Marc A. Deshusses. 2008. "The control of mercury vapor using biotrickling filters". *Chemosphere*, No. 3: 411–17. https://doi.org/10.1016/j.chemosphere.2007.06.073.

Pilon-Smits, Elizabeth A. H. 2005. "Phytoremediation". *The Annual Review of Plant Biology*, No. 56: 15–39. https://doi.org/10.1146/annurev.arplant.56.032604.144214.

Pratas, João, Carlos Paulo, Paulo J. C. Favas, and Venkatachalam Perumal. 2014. "Potential of aquatic plants for phytofiltration of uranium-contaminated waters in laboratory conditions". *Ecological Engineering*, No. 69: 170–6. https://doi.org/10.1016/j.ecoleng.2014.03.046.

Pujol, R., and S. Tarallo. 2000. "Total nitrogen removal in two-step biofiltration". *Water Science and Technology*, No. 4–5: 65–8. https://doi.org/10.2166/wst.2000.0427.

Ren, X., G. Zeng, L. Tang, J. Wang, J. Wan, J. Wang, Y. Deng, Y. Liu, and B. Peng. 2018. "The potential impact on the biodegradation of organic pollutants from composting technology for soil remediation". *Waste Management*, No. 72: 138–49. https://doi.org/10.1016/j.wasman.2017.11.032.

Rizwan, Muhammad, Shafaqat Ali, Muhammad Zia-ur-Rehman, Jörg Rinklebe, Daniel C. W. Tsang, Arooj Bashir, Arosha Maqbool, F. M. G. Tack, and Yong

Sik Ok. 2018. "Cadmium phytoremediation potential of Brassica crop species: A review". *Science of The Total Environment*, No. 631: 1175–91. https://doi.org/10.1016/j.scitotenv.2018.03.104.

Rosatto, Stefano, Enrica Roccotiello, Simone Di Piazza, Grazia Cecchi, Giuseppe Greco, Mirca Zotti, Luigi Vezzulli, and Mauro Mariotti. 2019. "Rhizosphere response to nickel in a facultative hyperaccumulator". *Chemosphere*, No. 232: 243–53. https://doi.org/10.1016/j.chemosphere.2019.05.193.

Salinas-Martnez, A., M. de los Santos-Cordova, O. Soto-Cruz, E. Delgado, H. Perez-Andrade, L. A. Hauad-Marroqun, and H. Medrano-Roldan. 2008. "Development of a bioremediation process by biostimulation of native microbial consortium through the heap leaching technique". *Journal of Environmental Management*, No. 1: 115–9. 10.1016/j.jenvman.2007.01.038.

Shahzad, Babar, Mohsin Tanveer, Abdul Rehman, Sardar Alam Cheema, Shah Fahad, Shams Ur Rehman, and Anket Sharma. 2018. "Nickel; whether toxic or essential for plants and environment – A review". *Plant Physiology and Biochemistry*, No. 132: 641–51. https://doi.org/10.1016/j.plaphy.2018.10.014.

Sharma, Sunita, Bikram Singh, and V. K. Manchanda. 2015. "Phytoremediation: Role of terrestrial plants and aquatic macrophytes in the remediation of radionuclides and heavy metal contaminated soil and water". *Environmental Science and Pollution Research International*, No. 2: 946–62. https://doi.org/10.1007/s11356-014-3635-8.

Singh, K., B. S. Giri, A. Sahi, S. R. Geed, M. K. Kureel, S. Singh, S. K. Dubey, B. N. Rai, S. Kumar, S. N. Upadhyay, and R. S. Singh. 2017. "Biofiltration of xylene using wood charcoal as the biofilter media under transient and high loading conditions". *Bioresource Technology*, No. 242: 351–8. https://doi.org/10.1016/j.biortech.2017.02.085.

Singh, Laxman K., G. Sudhakar, S. K. Swaminathan, and C. Muralidhar Rao. 2014. "Identification of elite native plants species for phytoaccumulation and remediation of major contaminants in uranium tailing ponds and its affected area". *Environment, Development and Sustainability*, No. 1: 57–81. https://doi.org/10.1007/s10668-014-9536-7.

Sychta, Klaudia, Aneta Słomka, Szymon Suski, Elżbieta Fiedor, Ewa L. Gregoraszczuk, and Elzbieta Kuta. 2018. "Suspended cells of metallicolous and nonmetallicolous *Viola* species tolerate, accumulate and detoxify zinc and lead". *Plant Physiology and Biochemistry*, No. 132: 666–74. https://doi.org/10.1016/j.plaphy.2018.10.013.

Tawussi, Frank, Dharmendra K. Gupta, Elena L. Mühr-Ebert, Stephanie Schneider, Stefan Bister, and Clemens Walther. 2017. "Uptake of Plutonium-238 into *Solanum tuberosum* L. (potato plants) in presence of complexing agent EDTA". *Journal of Environmental Radioactivity*, No. 178–179: 186–92. https://doi.org/10.1016/j.jenvrad.2017.08.007.

Tazdaït, Djaber, and Rym Salah-Tazdaït. 2021. "Polycyclic aromatic hydrocarbons: Toxicity and bioremediation approaches". In *Biotechnology for Sustainable Environment*, edited by Sanket J. Joshi, Arvind Deshmukh, and Hemen Sarma, 289–316. Singapore: Springer Nature. https://doi.org/10.1007/978-981-16-1955-7_12.

Tissut, Michel, Muriel Raveton, and Patrick Ravanel. 2006. "Ecoremediation. Cooperation between plants and soil microorganisms, molecular aspects and limits". In *Soil and Water Pollution Monitoring, Protection and Remediation*, edited by Irena Twardowska, Herbert E. Allen, Max M. Häggblom, and Sebastian Stefaniak, 489–504. Netherlands: Springer. https://doi.org/10.1007/978-1-4020-4728-2_32.

Vijayanand, K., P. Ganesh Prabu, A. Jaison Rathina Raj, and Anant Achary. 2012. "Studies on the bioremediation of chromium (vi) through bioleaching by *Thiobacillus ferrooxidans*". *International Journal of Research in Environmental Science and Technology*, No. 3: 54–60.

von der Heyden, Bjorn P., and Alakendra N. Roychoudhury. 2015. "Application, chemical interaction and fate of iron minerals in polluted sediment and soils". *Current Pollution Reports*, No. 1: 265–79. https://doi.org/10.1007/s40726-015-0020-2.

Wang, Dong, Er Nie, Xingzhang Luo, Xiaoying Yang, Qun Liu, and Zheng Zheng. 2015. "Study of nitrogen removal performance in pilot-scale multi-stage vermi-biofilter: Operating conditions impacts and nitrogen speciation transformation". *Environmental Earth Sciences*, No. 74: 3815–24. https://doi.org/10.1007/s12665-015-4713-z.

Watanabe, Kazuya. 2001. "Microorganisms relevant to bioremediation". *Current Opinion in Biotechnology*, No. 3: 237–41. https://doi.org/10.1016/S0958-1669(00)00205-6.

Weerasooriyagedara, Madara, Ahmed Ashiq, Anushka Upamali Rajapaksha, Rasika P. Wanigathunge, Tripti Agarwal, Dhammika Magana-Arachchi, and Meththika Vithanage. 2020. "Phytoremediation of fluoride from the environmental matrices: A review on its application strategies". *Groundwater for Sustainable Development*, No. 10. https://doi.org/10.1016/j.gsd.2020.100349.

Weis, Judith S., and Peddrick Weis. 2004. "Metal uptake, transport and release by wetland plants: Implications for phytoremediation and restoration". *Environment International*, No. 5: 685–700. https://doi.org/10.1016/j.envint.2003.11.002.

Wen, Qinxue, Qiong Wang, Xinqi Li, Zhiqiang Chen, Yingcai Tang, and Chongjian Zhang. 2018. "Enhanced organics and Cu^{2+} removal in electroplating wastewater by bioaugmentation". *Chemosphere*, No. 212: 476–485. https://doi.org/10.1016/j.chemosphere.2018.08.060.

Wong, Ming Hung. 2003. "Ecological restoration of mine degraded soils, with emphasis on metal contaminated soils". *Chemosphere*, No. 6: 775–80. https://doi.org/10.1016/S0045-6535(02)00232-1.

Wu, Manli, Warren A. Dick, Wei Li, Xiaochang Wang, Qian Yang, Tingting Wang, Limei Xu, Minghui Zhang, and Liming Chen. 2016. "Bioaugmentation and bio-stimulation of hydrocarbon degradation and the microbial community in a petroleum-contaminated soil". *International Biodeterioration & Biodegradation*, No. 107: 158–64. https://doi.org/10.1016/j.ibiod.2015.11.019.

Wu, Weijin, Xiaocui Liu, Xu Zhang, Minglong Zhu, and Wensong Tan. 2018. "Bioleaching of copper from waste printed circuit boards by bacteria-free cultural supernatant of iron – sulfur-oxidizing bacteria". *Bioresources and Bioprocessing*, No. 5. https://doi.org/10.1186/s40643-018-0196-6.

Xu, Jie, Harish Veeramani, Nikolla P. Qafoku, Gargi Singh, Maria V. Riquelme, Amy Pruden, Ravi K. Kukkadapu, Brandy N. Gartman, and Michael F. Hochella Jr. 2017. "Efficacy of acetate-amended biostimulation for uranium sequestration: Combined analysis of sediment/groundwater geochemistry and bacterial community structure". *Applied Geochemistry*, No. 78: 172–85. https://doi.org/10.1016/j.apgeochem.2016.12.024.

Yadav, Krishna Kumar, Neha Gupta, Amit Kumar, Lisa M. Reece, Neeraja Singh, Shahabaldin Rezania, and Shakeel A. Khan. 2018. "Mechanistic understanding and holistic approach of phytoremediation: A review on application and future prospects". *Ecological Engineering*, No. 120: 274–98. https://doi.org/10.1016/j.ecoleng.2018.05.039.

Yan, Huili, Gao Yiwei, Lulu Wu, Luyao Wang, Tian Zhang, Dai Changhua, Wenxiu Xu, Lu Feng, Mi Ma, Yong-Guan Zhu, and Zhenyan He. 2019. "Potential use of the *Pteris vittata* arsenic hyperaccumulation-regulation network for phytoremediation". *Journal of Hazardous Materials*, No. 368: 386–96. https://doi.org/10.1016/j.jhazmat.2019.01.072.

Yan, Yin, Yi Qun Zhou, and Cheng Hua Liang. 2015. "Evaluation of phosphate fertilizers for the immobilization of Cd in contaminated soils". *PLoS ONE*, No. 4: e0124022. https://doi.org/10.1371/journal.pone.0124022.

Zhu, Jian-Yu, Jing-Xia Zhang, Qian Li, Tao Han, Yue-Hua Hu, Xue-Duan Liu, Wen-Qing Qin, Li-Yuan Chai, and Guan-Zhou Qiu. 2014. "Bioleaching of heavy metals from contaminated alkaline sediment by auto- and heterotrophic bacteria in stirred tank reactor". *Transactions of Nonferrous Metals Society of China*, No. 9: 2969–75. https://doi.org/10.1016/S1003-6326(14)63433-6.

2 Radionuclides and heavy metals

Radionuclides and heavy metals: general considerations

Radionuclides

Elements can exist with a slightly different number of neutrons, and we call these isotopes of an element; this is particularly common for heavy elements where many neutrons are required to hold the nucleus together. In isotopes, there is the same number of protons, which means that the element is unchanged and has identical chemical properties. Much of the world around us comprises stable isotopes that do not disintegrate or have extremely long half-lives. However, in some cases, there are not enough neutrons in the nucleus or too many to be stable. In this case, the nucleus can spontaneously throw out particles (disintegrate) to reach a stable state. Isotopes that have unstable nuclei (excited nuclei or radionuclide) like this are known as radioactive isotopes. The more unstable a nucleus is, the faster it will reorganize itself into a more stable state; this process is called radioactive decay, during which three types of particles called α, β, and γ can arise over time. Alpha (α) decay has a positive charge, and it is a particle made up of two neutrons and two protons identical to the nucleus of the helium 4 atom ($_2^4He$) and a single sheet of paper can stop it. Beta (β) decay is an electron with a negative ($_{-1}^0\beta$) or a positive ($_{+1}^0\beta$) charge, and it can pass through 2 cm of Al plate but is stopped beyond this thickness. Gamma (γ) radiation has no charge at all with high energy (up to 20 GeV) (Salamon and Stecker 1998, 554). It is an extremely penetrating electromagnetic radiation capable of penetrating several centimetres of Pb or iron but is blocked by thick layers of concrete or Pb. Figure 2.1 gives some examples of reactions responsible for generating the three types of particles with the penetration characteristics of each of them.

It is generally considered that isotopes with an excess of neutron (^{14}C, ^{15}C, ^{23}Na, ^{137}Ba, etc.) tend to decay by emitting $_{-1}^0\beta$ particles (Baruah, Duorah,

DOI: 10.4324/9781003282600-3

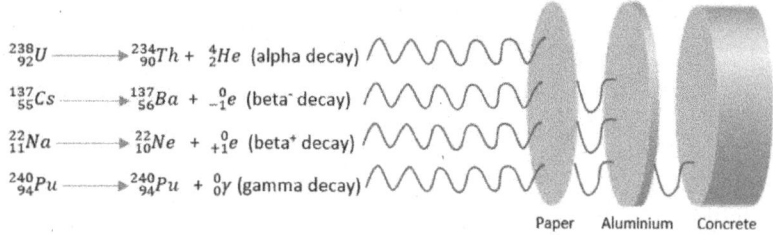

$$^{238}_{92}U \longrightarrow ^{234}_{90}Th + ^{4}_{2}He \text{ (alpha decay)}$$

$$^{137}_{55}Cs \longrightarrow ^{137}_{56}Ba + ^{0}_{-1}e \text{ (beta}^{-} \text{ decay)}$$

$$^{22}_{11}Na \longrightarrow ^{22}_{10}Ne + ^{0}_{+1}e \text{ (beta}^{+} \text{ decay)}$$

$$^{240}_{94}Pu \longrightarrow ^{240}_{94}Pu + ^{0}_{0}\gamma \text{ (gamma decay)}$$

Paper   Aluminium   Concrete

Figure 2.1 Types of radioactive decay: alpha, beta, and gamma with their respective penetration characteristics

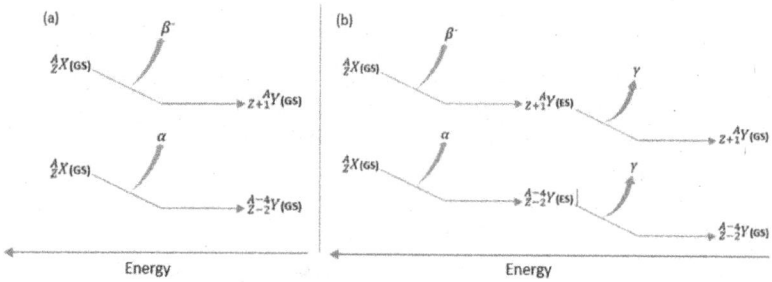

Figure 2.2 Pure beta and pure alpha decay are observed when both nuclei (mother and daughter) are in their GS (a). Beta and alpha decay associated with gamma radiation in the case the mother nucleus is in its GS and the daughter one in its ES (energy state above the ground state) (b).

and Duorah 2009, 174; L'Annunziata 2016, 168), while isotopes, which are deficient in neutrons (^{208}Ra, ^{203}At, ^{204}At, ^{209}At, etc.), decay by $^{0}_{+1}\beta$ particles emission, or by electron capture (no beta particle emission) (Devaraja, Beliuskina, Comas, Hofmann, et al. 2015, 201; L'Annunziata 2016, 174).

As depicted in Figure 2.2, alpha and beta decay can be pure or associated with the emission of one or more gamma rays. In pure beta decay, a conversion of a neutron to a proton occurs with negative beta particle emission, and because the mother and daughter nuclei are in their ground states (GS), there is no gamma emission during this decay process. This can be observed during ^{14}C to ^{14}N decay. When the mother and daughter nuclei are in their ground and excited states (ES), respectively, there is beta decay (β^{+} or β^{-}) associated with gamma emission. Similarly, the isotopes that undergo alpha emission do so as in beta decay described earlier.

The three different types of decay can occur naturally or anthropogenically. So far, 83 radionuclides (radioactive elements) are known in nature, including primordial (existed since the Earth creation) and cosmogenic (from extraterrestrial sources), and over 3,000 artificial (anthropogenic) radionuclides have been identified (L'Annunziata 2016, 68). The naturally occurring radiations result from spontaneous decay of isotopes of various elements present in nature, while the artificial radioactivity results from isotopes created artificially mainly from the fission of uranium-235 used as fuel to produce energy. The fission process is triggered by neutrons called slow neutrons with low energy and results in the production of another type of neutrons called fast neutrons (with high energy) along with isotopes of different elements, including certain new artificial radioactive nuclei such as Technetium-95 (Tc) or Promethium-147 (Pm) (not occurring on Earth) (Guillaumont 2019, 617).

Radioactive decay is a process that is entirely independent of any factor (chemical or physical) other than time (Pommé, Stroh, Altzitzoglou, Paepen, et al. 2018, 11), and the speed of the radioactive disintegration process is not constant. Besides, a given amount of radioactive element present at an instant t always requires the same period to decrease by half. This period is called half-life $(T_{1/2})$; it is specific to each radioactive isotope and ranges from microseconds (Polonium-214 (^{214}Po)) to billions of years (Potassium-40 (^{40}K), Uranium-238 (^{238}U)) (Waller 2012, 8474–5). Figure 2.3 shows this relationship graphically. The mother

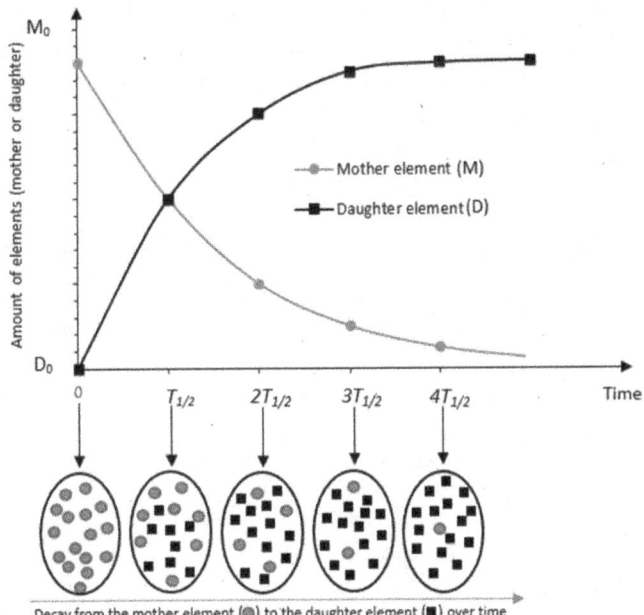

Figure 2.3 Evolution of the amount of mother and daughter elements over time.

element (M) quantity decreases exponentially over time; furthermore, its quantity at an instant t only depends on its quantity at t_0 and the time elapsed since t_0.

Another constant, called the disintegration constant (λ), is also a characteristic of each radionuclide; it is defined as the probability for a nucleus to decay into another element during a given period. The following relation relates half-life constant ($T_{1/2}$) and λ:

$$T_{1/2} = log 2 / \lambda \qquad (2.1)$$

The mathematical expression that expresses the radioactive decay of a given element is:

$$M_t = M_0 e^{-\lambda t} \qquad (2.2)$$

Where:

M_t and M_0 are the quantities of the mother element at times t and t_0, respectively.

Some examples of radionuclides produced in the environment, sorted by half-life, are presented in Table 2.1.

The units used to describe the decay of a radioactive element are all based on the determination of the quantity of this element that disintegrates per unit of time (disintegrations per second (dps), disintegrations per minute (dpm), disintegrations per hour (dph)). The Becquerel (Bq)

Table 2.1 Some characteristics of some radionuclides found in the environment

Mother element	Daughter element	Half-life (years)	Constant of disintegration (λ) (year^{-1})
^{238}U (Uranium)[a]	^{234}Th (Thorium)[a]	4.47×10^{9}[f]	1.55×10^{-11}
^{235}U (Uranium)[a]	^{231}Th (Thorium)[a]	0.70×10^{9}[f]	9.9×10^{-10}
^{129}I (Iodine)[b]	^{129}Xe (Xenon)[b]	1.57×10^{7}[b]	4.4×10^{-8}
^{99}Tc (Technetium)[c]	^{99}Ru (Ruthenium)[c]	2.1×10^{5}[c]	3.3×10^{-6}
^{237}Np (Neptunium)[a]	^{233}Pa (Protactinium)[a]	2.14×10^{6}[f]	3.2×10^{-7}
^{14}C (Carbon)[a]	^{14}N (Nitrogen)[a]	5715[a]	1.2×10^{-4}
^{32}Si (Silicon)[e]	^{32}P (Phosphorus)[e]	160[f]	4.3×10^{-3}

[b]Porcelli and Turekian 2003, 283; [c]Hidaka 2005, 249; [f]Lide 2010, 11(57), 11(189–90); [a]L'Annunziata 2016, 75, 77, 175; [e]Orrell, Arnquist, Bliss, Bunker, et al. 2018, 9

is the *Système International d'Unités* (SI) unit of radioactivity where by definition:

$$1 Becquerel\left(Bq\right) = 1 disintegration/sec\left(dps\right) \qquad (2.3)$$

One curie is, by definition, the activity of 1 g of pure Radium-226 (^{226}Ra).

$$1Ci = 3.7 \times 10^{10} dps \qquad (2.4)$$

In other words, $1Ci$ can be defined as the quantity of radionuclide giving 3.7×10^{10} *dps*.

Submultiples of the curie (millicurie (*mCi*), microcurie (*μCi*), and nanocurie (*nCi*)) can be used when the levels of radioactivity are very low (which is usually the case in the environment) and can be interrelated with the *dps* unit as follows:

$$1mCi = 3.7 \times 10^{7} dps \qquad (2.5)$$
$$1\mu Ci = 3.7 \times 10^{4} dps \qquad (2.6)$$
$$1nCi = 37 dps \qquad (2.7)$$

Another unit of radioactivity called rutherford (rd) is also used but less frequently. One rutherford is equivalent to 10^6 dps and 27×10^{-6} Ci (L'Annunziata 2016, 636–7).

The practical detection and measurement of the radioactivity in samples are performed using an instrument called the Geiger-Müller counter, which consists of two major parts, a detecting tube and a counting device. This monitoring instrument usually gives readouts in counts per second (cps), counts per minute (cpm), and counts per hour (dph) (Christ and Wernli Sr. 2014, 323). These counting rates can be related to the disintegration rates by a certain number of detection efficiency parameters like, for instance, the absolute efficiency of the detector (ε_{abs}) (Akkurt, Gunoglu, and Arda 2014, 1).

Two types of exposure to radionuclides exist. (1) Internal exposure occurs when the living beings absorb the radionuclides through inhalation, ingestion, and so forth; (2) external exposure is due to the deposition of materials carrying radionuclides on skin or clothes (IARC 2001, 31–2). To assess the exposure of the biological systems, especially the human being, to risk from radiations, some specific units have been defined, such as roentgen (R) (which quantifies exposure of living beings to ionizing radiations (γ rays or X-rays)). It is defined as the quantity of radiation necessary to generate one electrostatic unit of charge (esu) (3.33×10^{-10} C, which is equivalent to

2.1 × 10^9 ions) in one cm^3 of dry air at standard conditions of temperature and pressure (0°C, 760 mmHg) (National Research Council 1999, 274). Other units, such as radiation absorbed dose (rad), express the radiation dose absorbed by a living being instead. It is a unit, which quantifies the amount of radiation that produces 10^{-2} J of energy absorbed by 1 Kg of matter (tissue). The corresponding SI unit to the rad is the gray (Gy) (equivalent to 100 rad) (Magill and Galy 2005, 190). Another unit, which derives from the rad, is the erg per gram that expresses the quantity of radiation that liberates 10^{-4} J absorbed by one Kg of tissue. Therefore, this unit is equivalent to 10^{-2} rad and 10^{-4} Gy (National Research Council 1999, 273).

Lastly, given that the quantity of absorbed radiation is in no case an indication of its hazardous health effect on living beings, it would be useful to convert the dose of absorbed radiation into radiation dose equivalent, allowing the estimation of the health risk inherent to the absorbed radiation. The SI unit of radiation dose equivalent is the sievert (Sv), which measures the adverse health effect exerted by ionizing radiation on the biological system of the human being (Abe, Fuchigami, Hatamura, and Kasahara 2015, 140). It is the product of the absorbed dose measured in the unit gray by a corrective factor called radiation weighting (W_r) that depends on three parameters: the radiation type, radiation energy, and tissue being targeted (IARC 2001, 34).

Many applications are nowadays associated with the radioactive decay of elements, including medical applications in which a wide range of radioactive isotopes are used for diagnostic and therapeutic purposes. For instance, ^{99m}Tc, which is a 140 keV gamma-ray emitter with a short half-life of 6h, is by far the most largely used diagnostic agent for the identification of many diseases by classical or molecular imaging (Waller and Chowdhury 2016, 201; Ferreira, Lacerda, dos Santos, Branco de Barros, et al. 2017, 931; Miranda, Lemos, Fernandes, Ottoni, et al. 2019, 167). On the other hand, iodine-131, which has eight days half-life and is a 364.5 keV beta emitter, is mainly used as a radiopharmaceutical agent in many therapeutic applications (Neshasteh-Riz, Eyvazzadeh, Koosha, and Cheraghi 2017, 44; Fan, Wu, Lu, Yao, et al. 2019, 457). Some other examples of radioactive nuclides used in nuclear medicine therapy and diagnosis are listed in Table 2.2.

Toxicity

Nowadays, it is largely admitted that the benefits related to the applications of radionuclides in various fields, especially in health sciences, are undeniable. However, their hazardous effects on human health or the environment are well established and admitted as well. The therapeutic power of radiations is precisely related to their genotoxic properties (DNA deterioration) against diseased cells such as malignant tumour cells by killing them

Table 2.2 Physical characteristics of some common radionuclides used in nuclear medicine imaging and radiotherapy

Radionuclide	Half-life	Mode of decay	Application
[125]I	59.38 d	*EC, γ*	Therapeutic and diagnostic (in 't Hout, Schenk, van der Linden, and Roumen 2016, 12 ; Zheng, Wang, and Liu 2018, 317)
[64]Cu	12.7 h	*β⁻, β⁺, EC, γ*	Therapeutic and diagnostic (Wu, Chiu, Yu, Cautela, et al. 2018, 70)
[90]Y	2.66 d	*β⁻, γ*	Therapeutic (Kessler and Park 2019, 112)
[68]Ga	68 min	*β⁺, EC, γ*	Diagnostic (Nielsen, Kyneb, Alstrup, Jensen, et al. 2016, 603)
[67]Ga	3.26 d	*EC, γ*	Diagnostic (Wu, Chiu, Yu, Cautela, et al. 2018, 70)
[111]In	2.80 d	*EC, γ*	Diagnostic (Batchala, Dyer, Mukherjee, and Rehm 2019, 49)
[177]Lu	6.75 d	*β⁻, γ*	Therapeutic (D'Arienzo, Cozzella, Fazio, DeFelice, et al. 2016, 1745)

through senescence (cellular ageing), apoptosis (programmed cellular self-destruction), or necrosis (pathological cell death) (Wang, Wang, and Qian 2018, 2). The molecular DNA damages induced by exposure to ionizing radiations include single-strand breaks, in which the phosphate-deoxyribose backbone of one DNA strand is severed, double-strand breaks, in which the two strands of the DNA double helix are severed concomitantly, DNA cross-links, where a covalent link occurs between two nucleotides within the same strand (intrastrand crosslinks) or within the two complementary strands of the same DNA molecule (interstrand crosslinks), and non-enzymatic DNA-protein crosslinks, where there are non-specific covalent linkages between proteins and DNA (Stingele, Bellelli, and Boulton 2017, 3; Wang, Wang, and Qian 2018, 2). It is also well established that gamma rays induce various point mutations in human cells, including base substitutions (transitions, transversions), frameshifts, small deletions, and chromosome- and chromatid-type aberrations. Paradoxically, high doses of ionizing radiations used to treat cancer are also known to cause it. Therefore, it is essential to target the appropriate cells precisely during radiotherapy treatment and spare as much as possible the healthy cells. In the long term, the main health concern due to high radiation exposure is cancer development; target organs and cancer types depend on the type of radionuclides. It is generally considered that cancer is a process that is likely triggered by some six specific gene mutations happening within the same cell. Now, considering a mutation rate of ~10^{-7} per gene and per cell division (Blankenstein and Qin 2003, 187), it is unlikely that the whole cells of the body could accumulate such a high

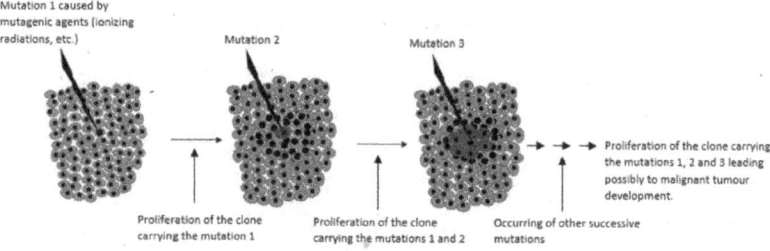

Figure 2.4 Progressive evolution of cancer

number of mutations (most of us would be dead). The probability of this happening in one of the 10^{14} cells of a person is very low (10^{14} x $(10^{-7})^6$ = 10^{-28}). Cancer occurs, however, owing to the combination of two mechanisms, namely an increase in the cell proliferation rates caused by certain mutations that generate a population of cells that may be targeted by another mutation (Figure 2.4) and a mutational phenomenon that affects the stability of the whole genome by increasing the global mutation rate.

There are three groups of genes whose mutation usually leads to cancer formation. (1) The oncogenes (*LMP-1, URG4/URGCP*, etc.) are the mutated versions of the proto-oncogenes, which normally positively regulate cell division. In contrary to the proto-oncogenes, the oncogenes are active in a way that is excessive or inappropriate (Dodurga, Seçme, and Şatıroğlu-Tufan 2018, 12). In the case of oncogenes, a single mutant allele may be sufficient to modify the phenotype of the cell. (2) The mutator genes such as *BRCA1* or *BRCA2* constitute a class of genes that play a fundamental role in DNA repair and replication; they are thus responsible for maintaining the integrity of the genome. The mutated versions of these genes lead to several cellular disorders, including malignant transformations (Clark and Pazdernik 2016, 611). (3) Tumor-suppressor genes or anti-oncogenes (*RB1, NF2, APC, DPC4*, etc.) are, as the name suggests, a group of genes that act by inhibiting tumour development. The two mutant alleles of a mutator gene or an anti-oncogene can lead to uncontrolled cell proliferation (Caldas and Venkitaraman 2001, 232–4).

The toxicity of radionuclides on human health, animals, and plants has been extensively studied for decades. Its extent depends on several factors, including the radionuclide type, the radiation type, the exposure time, the radiation energy, the target organism, and the pathway of intake (skin deposition, inhalation, ingestion, injection, etc.). Its manifestations (physiological, cellular, and molecular) are many and varied; they include acute signs, such as nausea, vomiting, radiation burns, and hair loss, or chronic signs, such as fatal or non-fatal cancers, and hereditary defects.

This chapter will describe the toxicity of two examples of radionuclides, namely, Radon-222 (^{222}Rn) and Plutonium-239 (^{239}Pu).

Radon-222 (^{222}Rn)

^{222}Rn (half-life of 3.8 days) is a naturally occurring radionuclide that decays sequentially over time by emitting α and β particles and produces a total of eight daughter radionuclides (^{218}Po, ^{214}Pb, ^{214}Bi, ^{214}Po, ^{210}Pb, ^{210}Bi, ^{210}Po and ^{206}Pb), of which the last-formed one (^{206}Pb) is stable (Gillmore, Crockett, and Przylibski 2010, 2052). Radon-222 is generated in the form of gas by the decay of Uranium-238 in rocks and building materials; it is one of the most abundant natural radionuclides in the environment and represents roughly half of the total human exposure to radiation. For instance, its air levels range from 4 to 20 Bq/m^3, while its concentration in naturally formed holes such as caves can reach 25,000 Bq/m^3 (Ojovan, Lee, and Kalmykov 2019, 44). Thus, ^{222}Rn can be a risk to human health at high levels. Furthermore, observations have demonstrated that gaseous ^{222}Rn has been associated with deletion mutations due to alpha-particle emission. According to WHO (2006), among the known risk factors for developing lung cancer, ^{222}Rn occupies the second place after smoking; it is responsible for tens of thousands of deaths worldwide every year. ^{222}Rn is incorporated in the body mainly through inhalation, and it deposits on different parts of the respiratory system (nasal cavity, larynx, trachea, bronchi, bronchioles, and alveoli), where it decays and acts on the target cells. It was estimated that the radiation absorbed dose of human bronchial basal epithelial cells ranged from 5 to 25 nGy. On the other hand, and based on epidemiological studies and physical dosimetry, it was determined that the radiation dose equivalent values of ^{222}Rn were ranging from 6 to 15 nSv (UNSCEAR 2000, 36). ^{222}Rn has also been positively associated with other non-cancer diseases: cardiovascular disease (coronary heart disease) (Villeneuve and Morrison 1997, 221).

Plutonium-239 (^{239}Pu)

^{239}Pu is considered one of the most toxic radionuclides to human health and the most concern causing ecological threats. This is because of its long half-life (24,100 years) and its high-energy alpha radiation produced during its decay to ^{235}U (Zhang and Hou 2019, 558). ^{239}Pu is an artificial fission product from nuclear reactions involving ^{238}U, which undergoes fission reaction by neutron capture giving, first ^{239}U, which will ultimately decay to ^{239}Pu (Hore-Lacy 2007, 48). The ^{239}Pu found in the environment springs mainly from the fallout of above-ground nuclear weapons testing in many regions of the world during the period 1940–60. Once released in the atmosphere, ^{239}Pu deposits on the ground in two forms (wet and dry). It is estimated that the average Pu concentrations in soil surfaces resulting from fallout vary from 0.01 to 0.1 pCi/g of soil, while the concentrations of ^{239}Pu in air vary between 1.6 x 10^{-6} and 3.8 x 10^{-6} pCi/m^3 of air (Rodriguez 2014, 985). ^{239}Pu constitutes thus undoubtedly a major health hazard for the populations living in the contaminated areas and the workers in the Pu production facilities who are permanently exposed, internally and externally, to a non-negligible risk of health problems. ^{239}Pu can be the source of acute and chronic toxicities. Acute toxicity occurs at high doses of ^{239}Pu over a short period of time; it is associated with many adverse effects, including metaplasia, pneumonitis, and fibrosis. Chronic toxicity results from low ^{239}Pu levels in a long period of time; it can lead to death very often by cancer. Many studies have been devoted to studying both acute and chronic toxicities of ^{239}Pu on many biological systems, including humans. For example, it has been demonstrated in a study by Shilnikova, Preston, Ron, Gilbert, et al. (2003, 791) that ^{239}Pu was responsible for 1,730 deaths due to solid cancers (lung, stomach, liver, skeleton, etc.) and 77 deaths due to leukaemia among workers in Mayak nuclear complex (the Russian Federation) exposed protractedly to high cumulative gamma radiation levels up to 10 Gy. In another study conducted by Gillies, Kuznetsova, Sokolnikov, Haylock, et al. (2017, 645), a clear relationship between cumulative internal exposure to ^{239}Pu and risk of death due to lung cancer was established among a cohort of 23,443 workers in Sellafield (former nuclear power generating site, UK).

Heavy metals

Heavy metals are metals with a high density (at least 5 g/cm^3) (Appenroth 2010, 21) and a high atomic mass (higher than 23) (Larrañaga, Lewis Sr., and Lewis 2016, 699). Brought back to the Earth with asteroids, heavy metals are found on the Earth's crust at low concentrations. Their presence in soils is mainly related to three natural origins: Earth's crust erosion, magma degassing during volcanic eruptions, and wildfires. Other processes involving human activities, like agro-industrial, domestic, automotive, medical, and electrical, contribute to extensive heavy metal distribution in specific polluted environments (landfills, stagnated wastewaters). Anthropogenic activities do not affect the global volume of heavy metals; however, they strongly influence their concentrations, chemical forms, and repartition (Li, Shen, Wai, and Li 2001, 215). It should be pointed out that, unlike natural heavy metals that are usually sequestered in relatively inert forms, those that originated anthropogenically are much reactive and thus much threatening to biological systems.

At trace levels, metal ions of Fe, Cu, Mn, Mo, and so on play a significant role in microorganisms and organisms' physiology as co-factors in many metabolic biochemical reactions involving enzymes called metalloenzymes

(Ni-Fe-hydrogenase, nitrile hydratase, sulphite oxidase, etc.). Metalloenzymes' activity strongly depends on the presence of metal ions, which can be directly bound to the protein part (the apoenzyme) of the enzyme or bound to its non-protein component called the prosthetic group (Hoppert 2011, 558–9). In addition, other proteins containing metal co-factors are involved in other physiological functions such as transport processes and storage. Another relevant role associated with one heavy metal in particular (Fe) is magnetotaxis, which is an active mode of motion found in a group of gram-negative bacteria called magnetotactic bacteria, involving specialized organelles, called magnetosomes, containing pure magnetite (Fe_3O_4) or greigite (Fe_3S_4) (Pósfai 2011, 538).

Toxicity

The mechanisms of toxicity of heavy metals are mainly based on the following three principles (Ochiai 1987, 361; Turna Demir and Yavuz 2020, 1): (1) inhibiting the enzymes by binding to the functional groups needed for their normal activities; (2) exerting genotoxic effects; (3) disorganization of cell membranes.

Heavy metals could easily reach humans through food and water since they can accumulate throughout the trophic chain. Many studies have demonstrated this, such as Milenkovic, Stajic, Stojic, Pucarevic, et al. (2019, 324), who assessed three toxic heavy metals (Cd, Hg, and Pb) in fish and seafood products collected from Serbian markets. The authors concluded that although heavy metal concentrations exceeded the maximum allowed values recommended by the European and Serbian regulation, frequent fish and seafood consumption may be harmful to health.

As stated before, some heavy metals, albeit toxic at high levels, are essential nutrients (required for various biochemical and physiological functions); others (Hg, Pb, As, Cd, and (Cr(VI)) are reported or at least suspected to be carcinogenic, mutagenic, teratogenic, allergenic, and endocrine-disrupting, even at very low concentrations (Rahman and Singh 2019, 1–2). Several biological systems have been used to study their behaviour towards the different chemical forms (elemental, organic, and inorganic) of toxic heavy metals. These include humans, animals, plants, and microorganisms. In this respect, in a study by Wani, Khan, and Zaidi (2008, 153), the effects of three heavy metals – namely, Cd, Cr, and Cu – were tested on some physiological parameters (growth, symbiosis, grain yield, and grain protein) of a pea. Among the three metals tested, Cu exhibited the most toxic effect on the plant by significantly decreasing seed yield, total dry matter, grain protein, and the number of nodules. In another study, the assessment of the effects of Ni, Cu, Co, Cd, Zn, and Pb at concentrations ranging from 10 to 100 mg/L on two marine bacterial enzymatic activities (catalase and dehydrogenase) was carried out (Kharchenko, Beleneva, Kovalchuk, and Hiep

2013, 289–93). All the metals decreased the activity of catalase in the tested bacteria (*Bacillus* sp. 1266, *Pseudomonas fluorescens* 1355), association 1 (*Bacillus* sp. 1259, *Bacillus* sp. 1266, *Bacillus* sp. 1393, *Pseudomonas fluorescens* 1355, and *Enterobacter* sp. 1360), association 2 (*Ruegeria* sp. 1444, *Pseudomonas putida* 1574, and an association of microorganisms from the seawater of Vostok Bay)). Cd and Cu presented the most important inhibitory effect, indicating their toxic effect on the studied bacteria. On the other hand, dehydrogenase activity in all microorganisms tested was completely inhibited in the presence of Cu at 50 and 100 mg/L. It should be noted that (1) catalase is an enzyme, which is known to play a crucial role in protecting cells against dangerous reactive oxygen molecules whose concentrations are increased in the presence of various toxic compounds, including heavy metals; (2) the bacterial dehydrogenases are enzymes that belong to the class of oxidoreductases and act by removing hydrogen from different natural or xenobiotic organic compounds.

There have been significant efforts in recent decades to study the toxic effects of heavy metals on human organisms, resulting in no doubt of their harmful effects on human health. Nowadays, Pb and Hg are the most preoccupying elements due to their high levels in the environment and their high toxicity. Table 2.3 depicts the adverse effects of some heavy metals on human physiology.

Table 2.3 Effects of some heavy metals on the human organism

Elements	Toxic forms	Effects	References
Mercury	Mercuric ion (Hg^{2+}) at 0.02 ppm	Acute toxicity on peripheral blood lymphocytes	McGuinness, Roess, and Barisas 1990, 131–9
	Mercuric ion (Hg^{2+}) at 7 µM	Toxicity against proximal tubular cells	Miles, Rodilla, Breen, Beattie, et al. 1999, 477–8
	Methyl mercury (MeHg)	It is positively associated with leukaemia	Yorifuji, Tsuda, and Kawakami 2007, 679
		It is positively associated with oxidative stress in autistic individuals	Leslie and Koger 2011, 313
Lead	Lead ion (Pb^{2+}) at 400 µM	Toxicity on dermal fibroblasts via DNA synthesis inhibition, intracellular ATP concentration depletion, and lysosomal membrane damage	Domínguez Solé, and Fortuny 2002, 49–51

Elements	Toxic forms	Effects	References
	Lead ion (Pb^{2+}) at 1 μM	Toxicity on erythrocytes by increasing potassium cation efflux, inhibiting sulphate influx, decreasing the reduced glutathione concentration, and cell surface damage	Gugliotta, De Luca, Romano, Rigano, et al. 2012, 590–2
	Lead ion (Pb^{2+}) at 3 mM	Toxicity on osteoblast-like cells via the inhibition of the calcium release activated calcium influx.	Jang, Kim, Kwon, Kim, et al. 2008, 190–1
Cadmium	Cadmium telluride (CdTe) quantum dots (particle diameters of 2.04, 3.24, and 5.4 nm)	Acute toxicity and potent carcinogenicity on bronchial epithelial cells	Zheng, Xu, Wu, Yao, et al. 2018, 899
	Cadmium ion (Cd^{2+}) at 20 and 30 μM	Apoptotic effect on astrocytes via increasing proapoptotic Bax expression and decreasing antiapoptotic Bcl-2 expression	Ospondpant, Phuagkhaopong, Suknuntha, Sangpairoj, et al. 2019, 55–6
	Cadmium ion (Cd^{2+}) at 10 μM	Premature oxidation of superoxide dismutase 1 in embryonic kidney cells	Polykretis, Cencetti, Donati, Luchinat, et al. 2019, 2–3
	Cadmium ion (Cd^{2+}) at 1–300 μM	Toxicity on intestinal epithelial cells via F-actin disorganization and via upregulating oxidative stress genes	Rusanov, Smirnova, Poromov, Fomicheva, et al. 2015, 1007–10
Chromium	Chromate (CrO_4^{2-}) at 25 and 50 μM	Toxicity on bronchial epithelial cells through reducing cell survival and inhibiting thioredoxin reductase activity	Myers and Myers 2009, 1477–8
	Dichromate ($Cr_2O_7^{2-}$) at 10–1000 μM	Toxicity on keratinocytes via reducing cell viability and mitochondrial activity	Curtis, Morton, Balafa, MacNeil, et al. 2007, 810–13
Nickel	Nickel ion (Ni^{2+}) at 75.5 μg/L	Toxicity on tongue, liver, and colon cells through reducing cell viability	Rincic Mlinaric, Durgo, Katic, and Spalj 2019, 2–4
	Nickel nanowires with diameter and length of 31 nm and 1 μm, respectively and a concentration of 118.8 μg/mL	Toxicity on fibroblasts via decreasing cell viability	Felix, Perez, Contreras, Ravasi, et al. 2016, 374–6

Biogeochemistry of radionuclides and heavy metals

This part of the chapter will review, in a concise manner, the current knowledge on the biogeochemistry, stability, and solubility of radionuclides and heavy metals in the environment.

Example of radionuclides: uranium

The environmental contamination by anthropogenic radionuclides has occurred relatively recently, since some decades ago. This contamination results from nuclear weapon tests and electronuclear power plant accidents, such as the most recent one in Fukushima Daiichi (Japan) in 2011, which has released vast quantities of radionuclides into the atmosphere. In addition to this, the nuclear industry generates large volumes of radioactive wastes, which are stored underground without treatment. Unfortunately, this disposal method is far from being safe since leaking from storage tanks poses a real threat to the environment by seriously contaminating surrounding soils and underground waters. Many leaks have been reported in many storage sites in the United States (Marshall, Beliaev, and Fredrickson 2010, 95). The hazard exhibited by these radioactive waste leaks is tightly linked to the chemical forms of their constituents. The oxidation state of the contaminant, its ability to interact either chemically by complexation with the geological organic or inorganic compounds (organic acids, HCO_3^-, SO_4^{2-}, F^-, HPO_4^{2-} and H_3SiO^{4-}) or biologically (biosorption, bioaccumulation) determines its chemical form, which in turn determines its environmental mobility. It is worth noting that pH and electrical potential (E_h) changes significantly influence the complexation/adsorption mechanisms (biological and chemical) (Pshinko 2009, 164–9; Sasaki, Zheng, Asano, and Kudo 2001, 221; Mahmoud, Rashad, Metwally, Saad, et al. 2017, 74).

In general, the more the radionuclides are oxidized, the more soluble they are in the aqueous medium, and inversely, the more they are reduced, the more they tend to precipitate. For example, uranium, whose natural abundance isotopes are 99.27% ^{238}U, 0.72% ^{235}U, and 0.005% ^{234}U (Sabol 2020, 2), can coexist in four different oxidation states, namely, U(II), U(IV), U(V), and U(VI), among which U(IV) and U(VI) are the most stable species in nature. In the aqueous environments, U(VI) and U(IV) form the chemical species uranyl (UO_2^{2+}) (soluble form) and uraninite (UO_2) (insoluble form), respectively (Marshall, Matthew, Alexander, Beliaev, et al. 2010, 96). Uranyl tends to form multiple soluble complexes with various anionic ligands, including carbonate, fluoride, phosphate, and organic compounds of humus. Besides, pH strongly influences the formation of hydrolysis species of uranyl; thus, in the presence of carbonates, neutral and higher pH values favour the formation of different uranium-carbonate species (($UO_2)_2CO_3(OH)_3^-$, $UO_2(CO_3)_2^{2-}$, $UO_2(CO_3)_3^{4-}$, etc.), while under carbonate-free conditions, UO_2^{2+} and various hydroxo-uranyl complex species appear depending on the pH of the medium. At pH < 6, UO_2^{2+}, UO_2OH^+, $UO_2(OH)_2$, $(UO_2)_2(OH)_2^{2+}$, $(UO_2)_3(OH)_5^+$, $(UO_2)_4(OH)_7^+$, $(UO_2)_4(OH)_2^{6+}$, and so on dominate, while at neutral and higher pH $UO_2(OH)^{3-}$, $(UO_2)_3(OH)_8^{2-}$, $(UO_2)_3(OH)_{10}^{4-}$, and so on tend to prevail (Krestou and Panias 2004, 118–25).

Example of heavy metals: mercury

Like other materials such as oxygen, carbon, and nitrogen, heavy metals are subjected to a natural circulation through different spheres of the Earth – namely, the lithosphere, biosphere, atmosphere, cryosphere, and hydrosphere. The biogeochemical cycle is a pathway by which a given element travels through these spheres. It is tightly linked to the concentration, dissemination, and chemical modifications of the element within the same sphere and its dynamic rate between different spheres. The chemical speciation of the element in the different spheres of the cycle is essential to understand appropriately, describe, and eventually predict its behaviour in the environment (dispersion, sinking, toxicity).

This part of the chapter will focus on one example of heavy metals (Hg) that is reported to be a persistent widespread environmental and industrial pollutant susceptible to cause enormous human health impairments (Bhan and Sarkar 2005, 39). Hg is a ubiquitous and persistent heavy metal that occurs naturally in the environment in three chemical forms, elemental (metallic) (Hg^0), mercurous (Hg_2^{2+}), and mercuric (Hg^{2+}). The metallic form of Hg is liquid at room temperature, but because of its high vapour pressure, it becomes vapour at higher temperatures and ends up in the atmosphere as mercury vapour. Besides, gaseous Hg^0 can move from the water surface of aquatic ecosystems (hydrosphere) to the atmosphere essentially through the process of converting Hg^{2+} to Hg^0 by photoproduction (Mason, Fitzgerald, and Morel 1994, 3192).

Hg^{2+} and Hg^0 can interact with carbon to form organic Hg compounds such as monomethylmercury, dimethylmercury, ethylmercury, and phenylmercury. Monomethylmercury (MeHg) is the most frequently encountered in the environment and can be easily accumulated in the tissues of organisms, including humans, causing severe toxic effects. MeHg can result from both abiotic and biotic processes. The biotic process is dominant in the environment and occurs by the action of soil and aquatic microorganisms, mainly in anaerobic conditions. This process consists mainly of transferring a methyl group from a donor molecule (methylcobalamin) to mercuric cation through the acetyl-coenzyme A pathway (Ma, Du, and Wang 2019,

1901). It is worth noting that Hg^0 can be formed from MeHg (Lehnherr and St. Louis 2009, 5692) or Hg^{2+} (Amyot, Mierle, Lean, and McQueen 1994, 2366) through abiotic photoreduction, which occurs in aquatic media.

Moreover, Hg in its divalent forms (Hg^{2+} and Hg_2^{2+}) is susceptible to combine with several inorganic elements such as oxygen, chlorine, and sulphur to form complexes, which are very soluble in water and could easily and significantly move from the areas that they are contaminating to uncontaminated ones. On the other hand, Hg^{2+} is the predominant form found in aquatic media, while Hg_2^{2+} is relatively negligible in such media. The formation of Hg^{2+} in water is observed in the presence of different oxidizing compounds, including chloride, bromide, and organic acids (Batrakova, Travnikov, and Rozovskaya 2014, 1051).

Reference list

Abe, S., M. Fuchigami, Y. Hatamura, and N. Kasahara. 2015. *The 2011 Fukushima Nuclear Power Plant Accident, How and Why It Happened*. Cambridge: Woodhead Publishing.

Akkurt, I., K. Gunoglu, and S. S. Arda. 2014. "Detection efficiency of NaI(Tl) detector in 511–1332 keV energy range". *Science and Technology of Nuclear Installations*, No. 2014: 1–5. https://doi.org/10.1155/2014/186798.

Amyot, Marc, Greg Mierle, David R. S. Lean, and Donald J. McQueen. 1994. "Sunlight-induced formation of dissolved gaseous mercury in lake waters". *Environmental Science and Technology*, No. 13: 2366–71. https://doi.org/10.1021/es00062a022.

Appenroth, Klaus-J. 2010. "Definition of "heavy metals" and their role in biological systems". In *Soil Heavy Metals*, edited by Irena Sherameti, and Ajit Varma, 19–29. Heidelberg: Springer.

Baruah, Rulee, Kalpana Duorah, K., and H. L. Duorah. 2009. "Rapid neutron capture process in supernovae and chemical element formation". *Journal of Astrophysics and Astronomy*, No. 3–4: 165–75. https://doi.org/10.1007/s12036-009-0013-x.

Batchala, Prem P., Anthony Dyer, Sugoto Mukherjee, and Patrice K. Rehm. 2019. "Lateral ectopic thyroid mimics carotid body tumor on Indium-111 pentetreotide scintigraphy". *Clinical Imaging*, No. 58: 46–9. https://doi.org/10.1016/j.clinimag.2019.05.012.

Batrakova, N., O. Travnikov, and O. Rozovskaya. 2014 "Chemical and physical transformations of mercury in the ocean: A review". *Ocean Science*, No. 10: 1047–63. https://doi.org/10.5194/os-10-1047-2014.

Bhan, Ashima, and N. N. Sarkar. 2005. "Mercury in the environment: Effects on health and reproduction". *Reviews on Environmental Health*, No. 1: 39–56. https://doi.org/10.1515/REVEH.2005.20.1.39.

Blankenstein, Thomas, and Zhihai Qin. 2003. "Chemical carcinogens as foreign bodies and some pitfalls regarding cancer immune surveillance". *Advances in Cancer Research*, No. 90: 179–207. https://doi.org/10.1016/S0065-230X(03)90006-6.

Caldas, C., and A. R. Venkitaraman (2001) "Tumor suppressor genes". *Brenner's Encyclopedia of Genetics*, No. 7: 232–7. https://doi.org/10.1016/B978-0-12-374984-0.01595-3.

Christ, Robert D., and Robert L. Wernli Sr. 2014. *The ROV Manual: A User Guide for Remotely Operated Vehicles*. Waltham: Butterworth Heinemann.

Clark, David P., and Nanette J. Pazdernik. 2016. *Biotechnology*. London: Academic Cell.

Curtis, Angela, Jackie Morton, Chariklia Balafa, Sheila MacNeil, David J. Gawkrodger, Nicholas D. Warren, and Gareth S. Evans. 2007. "The effects of nickel and chromium on human keratinocytes: Differences in viability, cell associated metal and IL-1a release". *Toxicology in Vitro*, No. 5: 809–19. https://doi.org/10.1016/j.tiv.2007.01.026.

D'Arienzo, M., M. L. Cozzella, A. Fazio, P. DeFelice, G. Iaccarino, M. D'Andrea, S. Ungania, M. Cazzato, K. Schmidt, S. Kimiaei, and L. Strigari. 2016. "Quantitative ^{177}Lu SPECT imaging using advanced correction algorithms in non-reference geometry". *European Journal of Medical Physics*, No. 12: 1745–52. https://doi.org/10.1016/j.ejmp.2016.09.014.

Devaraja, H., M. S. Heinz, O. Beliuskina, V. Comas, S. Hofmann, C. Hornung, G. Münzenberg, K. Nishio, D. Ackermann, Y. K. Gambhir, M. Gupta, R. A. Henderson, F. P. Heßberger, J. Khuyagbaatar, B. Kindler, B. Lommel, K. J. Moody, J. Maurer, R. Mann, A. G. Popeko, D. A. Shaughnessy, M. A. Stoyer, and A. V. Yeremin. 2015. "Observation of new neutron-deficient isotopes with Z ≥ 92 in multinucleon transfer reactions". *Physics Letters B*, No. 748: 199–203. https://doi.org/10.1016/j.physletb.2015.07.006.

Dodurga, Yavuz, Mücahit Seçme, and N. Lale Şatıroğlu-Tufan. 2018. "A novel oncogene URG4/URGCP and its role in cancer". *Gene*, No. 668: 12–17. https://doi.org/10.1016/j.gene.2018.05.047.

Domínguez, C., E. Solé, and A. Fortuny. 2002. "In vitro lead-induced cell toxicity and cytoprotective activity of fetal calf serum in human fibroblasts". *Molecular and Cellular Biochemistry*, No. 237: 47–53. https://doi.org/10.1023/A:1016547519763.

Fan, Wenzhe, Yanqin Wu, Mingjian Lu, Wang Yao, Wei Cui, Yue Zhao, Yu Wang, and Jiaping Li. 2019. "A meta-analysis of the efficacy and safety of iodine [^{131}I] metuximab infusion combined with TACE for treatment of hepatocellular carcinoma". *Clinics and Research in Hepatology and Gastroenterology*, No. 4: 451–45. https://doi.org/10.1016/j.clinre.2018.09.006.

Felix, Laura P., Jose E. Perez, Maria F. Contreras, Timothy Ravasi, and Jürgen Kosel. 2016. "Cytotoxic effects of nickel nanowires in human fibroblasts". *Toxicology Reports*, No. 3: 373–80. http://doi.org/10.1016/j.toxrep.2016.03.004.

Ferreira, Iêda Mendes, Camila Maria de Sousa Lacerda, Sara Roberta dos Santos, André Luís Branco de Barros, Simone Odília Fernandes, Valbert Nascimento Cardoso, and Antero Silva Ribeiro de Andrade. 2017. "Detection of bacterial infection by a technetium-99m-labeled peptidoglycan aptamer". *Biomedicine and Pharmacotherapy*, No. 93: 931–8. http://doi.org/10.1016/j.biopha.2017.07.017.

Gillies, Michael, Irina Kuznetsova, Mikhail Sokolnikov, Richard Haylock, Jackie O'Hagan, Yulia Tsareva, and Elena Labutina. 2017. "Lung cancer risk from plutonium: A pooled analysis of the Mayak and Sellafield worker cohorts". *Radiation Research*, No. 6: 725–40. https://doi.org/10.1667/RR14719.1.

Gillmore, G. K., R. G. M. Crockett, and T. A. Przylibski. 2010. "IGCP Project 571: Radon, health and natural hazards". *Natural Hazards and Earth System Sciences*, No. 10: 2051–4. https://doi.org/10.5194/nhess-10-2051-2010.

Gugliotta, Tiziana, Grazia De Luca, Pietro Romano, Caterina Rigano, Adriana Scuteri, and Leonardo Romano. 2012. "Effects of lead chloride on human erythrocyte membranes and on kinetic anion sulphate and glutathione concentrations". *Cellular and Molecular Biology Letters*, No. 4: 586–97. https://doi.org/10.2478/s11658-012-0027-2.

Guillaumont, Robert. 2019. "Completion and extension of the periodic table of elements beyond uranium". *Comptes Rendus Physique*, No. 7–8: 617–30. https://doi.org/10.1016/j.crhy.2018.12.006.

Hidaka, Hiroshi. 2005. "Technetium in cosmo- and geochemical fields". *Journal of Nuclear and Radiochemical Sciences*, No. 3: 249–52.

Hoppert, Michael. 2011. "Metalloenzymes". In *Encyclopedia of Geobiology. Encyclopedia of Earth Sciences Series*, edited by Joachim Reitner, and Volker Thiel, 558–63. Dordrecht: Springer. https://doi.org/10.1007/978-1-4020-9212-1_134.

Hore-Lacy, Ian. 2007. *Nuclear Energy in the 21st Century*. Cambridge, MA: Academic Press.

International Agency for Research on Cancer (IARC). 2001. "IARC Monographs on the Evaluation of Carcinogenic Risks to Humans Ionizing Radiation, Part 2: Some Internally Deposited Radionuclides". Lyon: IARC Press. 78. 1–563.

in 't Hout, B. A., K. E. Schenk, A. N. van der Linden, and R. M. H. Roumen. 2016. "Efficacy of localization of non-palpable, invasive breast cancer: Wire localization vs. Iodine-125 seed: A historical comparison". *The Breast*, No. 29: 8–13. https://doi.org/10.1016/j.breast.2016.06.011.

Jang, Hye-Ock, Ji-Suk Kim, Woo-Cheol Kwon, Jeong-Kuk Kim, Myung-Suk Ko, Dong-Hoo Kim, Won-Il Kim, Young-Chan Jeon, In-Kyo Chung, Sang-Hun Shin, Jin Chung, Moon-Kyung Bae, and Il Yun. 2008. "The effect of lead on calcium release activated calcium influx in primary cultures of human osteoblast-like cells". *Archives of Pharmacal Research*, No. 2: 188–94. https://doi.org/10.1007/s12272-001-1140-3.

Kessler, Jonathan, and John J. Park. 2019. "Yttrium-90 Radioembolization after local hepatic therapy: How prior treatments impact patient selection, dosing, and toxicity". *Techniques in Vascular and Interventional Radiology*, No. 2: 112–16. https://doi.org/10.1053/j.tvir.2019.02.012.

Kharchenko, U. V., I. A. Beleneva, Yu. L. Kovalchuk, and L. T. M. Hiep. 2013. "Enzymatic indication of heavy metal toxicity to marine heterotrophic bacteria". *Russian Journal of Marine Biology*, No. 4: 287–94. https://doi.org/10.1134/S1063074013040068.

Krestou, A., and Dimitris Panias. 2004. "Uranium (VI) speciation diagrams in the $UO_2^{2+}/CO_3^{2-}/H_2O$ System at 25°C". *The European Journal of Mineral Processing and Environmental Protection*, No. 2: 113–29.

L'Annunziata, Michael F. 2016. *Radioactivity: Introduction and History, from the Quantum to Quarks*. Amsterdam: Elsevier B.V.

Larrañaga, Michael D., Richard J. Lewis Sr., and Robert A. Lewis. 2016. *Hawley's Condensed Chemical Dictionary*. Hoboken: Wiley.

Lehnherr, Igor, and Vincent l. St. Louis. 2009. "Importance of ultraviolet radiation in the photodemethylation of methylmercury in freshwater ecosystems". *Environmental Science and Technology*, No. 15: 5692–8. https://doi.org/10.1021/es9002923.

Leslie, Kerry E., and Susan M. Koger. 2011. "A significant factor in autism: Methyl mercury induced oxidative stress in genetically susceptible individuals". *Journal of Developmental and Physical Disabilities*, No. 3: 313–24. https://doi.org/10.1007/s10882-011-9230-8.

Li, Xiangdong, Zhenguo Shen, Onyx W. H Wai, and Yok-Sheung Li. 2001. "Chemical forms of Pb, Zn and Cu in the sediment profiles of the Pearl River Estuary". *Marine Pollution Bulletin*, No. 3: 215–23. https://doi.org/10.1016/S0025-326X(00)00145-4.

Lide, David R., ed. 2010. *CRC Handbook of Chemistry and Physics*. Boca Raton: CRC Press/Taylor and Francis.

Ma, Ming, Hongxia Du, and Dingyong Wang. 2019. "Mercury methylation by anaerobic microorganisms: A review". *Critical Reviews in Environmental Science and Technology*, No. 20: 1893–936. https://doi.org/10.1080/10643389.2019.1594517.

Magill, Joseph, and Jean Galy. 2005. *Radioactivity Radionuclides Radiation*. Berlin Heidelberg: Springer-Verlag.

Mahmoud Mamdoh R., Ghada M. Rashad, Essam Metwally, Ebtissam A. Saad, and Ahmed M. Elewa. 2017. "Adsorptive removal of radionuclides from aqueous solutions using sepiolite: Single and multi-component systems". *Applied Clay Science*, No. 141: 72–80. https://doi.org/10.1016/j.clay.2016.12.021.

Marshall, Matthew J., Alexander S. Beliaev, and James K. Fredrickson. 2010. "Microbiological transformations of radionuclides in the subsurface". In *Environmental Microbiology*, edited by Ralph Mitchell, and Ji-Dong Gu, 95–114. Hoboken: Wiley-Blackwell.

Mason, Robert P., W. F. Fitzgerald, and François M. M. Morel. 1994. "The biogeochemical cycling of elemental mercury: Anthropogenic influences". *Geochimica et Cosmochimica Acta*, No. 15: 3191–319. http://doi.org/10.1016/0016-7037(94)90046-9.

McGuinness, Susan M., Deborah A. Roess, and B. George Barisas. 1990. "Acute toxicity effects of mercury and other heavy metals on HeLa cells and human lymphocytes evaluated via microcalorimetry". *Thermochimica Acta*, No. 172: 131–45. https://doi.org/10.1016/0040-6031(90)80567-I.

Milenkovic, Biljana, Jelena M. Stajic, Natasa Stojic, Mira Pucarevic, and Snezana Strbac. 2019. "Evaluation of heavy metals and radionuclides in fish and seafood products". *Chemosphere*, No. 229: 324–31. https://doi.org/10.1016/j.chemosphere.2019.04.189.

Miles, Adrian T., Vicente Rodilla, Antony G. Breen, John Beattie Z., William Jenner, and Gabrielle M. Hawksworth. 1999. "Metallothionein induction in human proximal tubular cell cultures – Lack of protection against heavy metal toxicity". In *Metallothionein IV*, edited by C. Klaassen, 477–84. Basel/Switzerland: Birkhauser Verlag. https://doi.org/10.1007/978-3-0348-8847-9_70.

Miranda, Sued E. M., Janaína A. Lemos, Renata S. Fernandes, Flaviano Melo Ottoni, Ricardo J. Alves, Alice Ferretti, Domenico Rubello, Valbert N. Cardoso, and

André L. B. de Barros. 2019. "Technetium-99m-labeled lapachol as an imaging probe for breast tumor identification". *Revista Española de Medicina Nuclear e Imagen Molecular*, No. 3: 167–72. https://doi.org/10.1016/j.remnie.2018.11.002.

Myers, Judith M., and Charles R. Myers. 2009. "The effects of hexavalent chromium on thioredoxin reductase and peroxiredoxins in human bronchial epithelial cells". *Free Radical Biology and Medicine*, No. 10: 1477–85. https://doi.org/10.1016/j.freeradbiomed.2009.08.015.

National Research Council. 1999. *Evaluation of Guidelines for Exposures to Technologically Enhanced Naturally Occurring Radioactive Materials*. Washington, DC: The National Academies Press. https://doi.org/10.17226/6360.

Neshasteh-Riz, Ali, Nazila Eyvazzadeh, Fereshteh Koosha, and Susan Cheraghi. 2017. "Comparison of DSB effects of the beta particles of iodine-131 and 6 MV X-Ray at a dose of 2 Gy in the presence of 2-methoxyestradiol, IUdR, and TPT in glioblastoma spheroids". *Radiation Physics and Chemistry*, No. 131: 41–5. https://doi.org/10.1016/j.radphyschem.2016.10.011.

Nielsen, Karin M., Majbritt H. Kyneb, Aage K. O. Alstrup, Jakob J. Jensen, Dirk Bender, Henrik C. Schønheyder, Pia Afzelius, Ole L. Nielsen, and Svend B. Jensen. 2016. "[68]Ga-labeled phage-display selected peptides as tracers for positron emission tomography imaging of *Staphylococcus aureus* biofilm-associated infections: Selection, radiolabelling and preliminary biological evaluation". *Nuclear Medicine and Biology*, No. 10: 593–605. http://doi.org/10.1016/j.nucmedbio.2016.07.002.

Ochiai, Ei-Ichiro. 1987. *General Principles of Biochemistry of the Elements*. New York: Plenum Press.

Ojovan, Michael I., William E., Lee, and Stepan N., Kalmykov. 2019. *An Introduction to Nuclear Waste Immobilisation*. Amsterdam: Elsevier.

Orrell, John L., Isaac J. Arnquist, Mary Bliss, Raymond Bunker, and Zachary S. Finch. 2018. "Naturally occurring 32Si and low-background silicon dark matter detectors". *Astroparticle Physics*, No. 99: 9–20. https://doi.org/10.1016/j.astropartphys.2018.02.005.

Ospondpant, Dusadee, Suttinee Phuagkhaopong, Kran Suknuntha, Kant Sangpairoj, Thitima Kasemsuk, Chutima Srimaroeng, and Pornpun Vivithanaporn. 2019. "Cadmium induces apoptotic program imbalance and cell cycle inhibitor expression in cultured human astrocytes". *Environmental Toxicology and Pharmacology*, No. 65: 53–9. https://doi.org/10.1016/j.etap.2018.12.001.

Polykretis, Panagis, Francesca Cencetti, Chiara Donati, Enrico Luchinat, and Lucia Banci. 2019. "Cadmium effects on superoxide dismutase 1 in human cells revealed by NMR". *Redox Biology*, No. 21: 1–7. https://doi.org/10.1016/j.redox.2019.101102.

Pommé, S., H. Stroh, T. Altzitzoglou, J. Paepen, R. Van Ammel, K. Kossert, O. Nähle, J. D. Keightley, K. M. Ferreira, L. Verheyen, and M. Bruggeman. 2018. "Is decay constant?". *Applied Radiation and Isotopes*, No. 134: 6–12. https://doi.org/10.1016/j.apradiso.2017.09.002.

Porcelli, D., and K. K. Turekian. 2003. "The history of planetary degassing as recorded by noble gases". *In Treatise on Geochemistry*, edited by Heinrich D. Holland and Karl K. Turekian, 281–318. Amsterdam: Elsevier. https://doi.org/10.1016/B0-08-043751-6/04181-5.

Pósfai, Mihály. 2011. "Magnetotactic bacteria". In *Encyclopedia of Geobiology. Encyclopedia of Earth Sciences Series*, edited by Joachim Reitner and Volker Thiel, 537–41. Dordrecht: Springer. https://doi.org/10.1007/978-1-4020-9212-1.

Pshinko, G. N. 2009. "Impact of humic matter on sorption of radionuclides by mont-morrilonite". *Journal of Water Chemistry and Technology*, No. 3: 163–71. https://doi.org/10.3103/S1063455X09030047.

Rahman, Zeeshanur, and Ved Pal Singh. 2019. "The relative impact of toxic heavy metals (THMs) (arsenic (As), Cadmium (Cd), Chromium (Cr)(VI), Mercury (Hg), and lead (Pb)) on the total environment: An overview". *Environmental Monitoring and Assessment*, No. 7: 419. https://doi.org/10.1007/s10661-019-7528-7.

Rincic Mlinaric, M., K. Durgo, V. Katic, and S. Spalj. 2019. "Cytotoxicity and oxidative stress induced by nickel and titanium ions from dental alloys on cells of gastrointestinal tract". *Toxicology and Applied Pharmacology*, No. 383: 1–10. https://doi.org/10.1016/j.taap.2019.114784.

Rodriguez, Y. R. 2014. "Plutonium". In *Encyclopedia of Toxicology*, edited by Philip Wexler, 982–5. Cambridge, MA: Academic Press.

Rusanov, A. L., A. V. Smirnova, A. A. Poromov, K. A. Fomicheva, N. G. Luzgina, and A. G. Majouga. 2015. "Effects of cadmium chloride on the functional state of human intestinal cells". *Toxicology in Vitro*, No. 5: 1006–11. http://doi.org/10.1016/j.tiv.2015.03.018.

Sabol, Jozef. 2020. "Uranium in the beginning of the nuclear age: Reflections on the historical role of jáchymov and an overview of early and present epidemiological studies". In *Radionuclides and Heavy Metals in the Environment*, edited by Dharmendra K. Gupta and Clemens Walther, 1–32. Cham, Switzerland: Springer Nature.

Salamon M. H., and F. W. Stecker. 1998. "Absorption of high-energy gamma rays by interactions with extragalactic starlight photons at high redshifts and the high-energy gamma-ray background". *The Astrophysical Journal*. No. 2: 547–54.

Sasaki, Takayuki, James Zheng, Hidekazu Asano, and Akira Kudo. 2001. "Interaction of Pu, Np and Pa with anaerobic microorganisms at geologic repositories". *Radioactivity in the Environment*, No. 1: 221–32. https://doi.org/10.1016/S1569-4860(01)80016-4.

Shilnikova, N. S., D. L. Preston, E. Ron, E. S. Gilbert, E. K. Vassilenko, S. A. Romanov, I. S. Kuznetsova, M. E. Sokolnikov, P. V. Okatenko, V. V. Kreslov, and N. A. Koshurnikova. 2003. "Cancer mortality risk among workers at the Mayak Nuclear Complex". *Radiation Research*, No. 6: 787–98. https://doi.org/10.1667/0 0337587(2003)159[0787:CMRAWA]2.0.CO;2.

Stingele, Julian, Roberto Bellelli, and Simon J. Boulton. 2017. "Mechanisms of DNA-protein crosslink repair". *Nature Reviews Molecular Cell Biology*, No. 18: 563–73. https://doi.org/10.1038/nrm.2017.56.

Turna Demir, Fatma, and Mustafa Yavuz. 2020. "Heavy metal accumulation and genotoxic effects in Levant Vole (*Microtus guentheri*) collected from contaminated areas due to mining activities". *Environmental Pollution*, No. 256: 113378. https://doi.org/10.1016/j.envpol.2019.113378.

United Nations Scientific Committee on the Effects of Atomic Radiation (UNSCEAR). 2000. "Sources and effects of ionizing radiation, Annex B: Exposures from natural radiation sources". *United Nations Publication*, No. 1: 83–156. www.unscear.org/docs/publications/2000/UNSCEAR_2000_Report_Vol.I.pdf.

Villeneuve, P. J., and H. I. Morrison. 1997. "Coronary heart disease mortality among Newfoundland fluorspar miners". *Scandinavian Journal of Work Environment and Health*, No. 3: 221–6. https://doi.org/10.5271/sjweh.202.

Waller, Edward J. 2012. "Radiation in the environment, sources of". In *Encyclopedia of sustainability science and technology*, edited by Robert A. Meyers, 8467–8536. New York: Springer. https://doi.org/10.1007/978-1-4419-0851-3_323.

Waller, Michael L., and Fahmid U. Chowdhury. 2016. "The basic science of nuclear medicine". *Orthopaedics and Trauma*, No. 3: 201–22. https://doi.org/10.1016/j.mporth.2016.05.013.

Wang, Jin-song, Hai-juan Wang, and Hai-Li Qian. 2018. "Biological effects of radiation on cancer cells". *Military Medical Research*, No. 20: 1–10. https://doi.org/10.1186/s40779-018-0167-4.

Wani, Parvaze Ahmad, Mohammad Saghir Khan, and Almas Zaidi. 2008. "Effects of heavy metal toxicity on growth, symbiosis, seed yield and metal uptake in pea grown in metal amended soil". *Bulletin of Environmental Contamination and Toxicology*, No. 81: 152–8. https://doi.org/10.1007/s00128-008-9383-z.

World Health Organization (WHO). 2006. "Improve Home Ventilation to Reduce Radon Levels, WHO Warns". Accessed September 13, 2019. www.who.int/mediacentre/news/notes/2006/np02/en/.

Wu, Tsai-Jung, Hsiao-Yu Chiu, John Yu, Mafalda P. Cautela, Bruno Sarmento, José das Neves, Carme Catala, Nicolas Pazos-Perez, Luca Guerrini, Ramon A. Alvarez-Puebla, Sanja Vranješ-Đurić, and Nenad L. Ignjatović. 2018. "Nanotechnologies for early diagnosis, *in situ* disease monitoring, and prevention". In *Nanotechnologies in Preventive and Regenerative Medicine*, edited by Vuk Uskoković and Dragan P. Uskoković, 1–92. Amsterdam: Elsevier. https://doi.org/10.1016/B978-0-323-48063-5.00001-0.

Yorifuji, Takashi, Toshihide Tsuda, and Norito Kawakami. 2007. "Age standardized cancer mortality ratios in areas heavily exposed to methyl mercury". *International Archives of Occupational and Environmental Health*, No. 80: 679–88. https://doi.org/10.1007/s00420-007-0179-y.

Zhang, Weichao, and Xiaolin Hou. 2019. "Level, distribution and sources of plutonium in the coastal areas of China". *Chemosphere*, No. 230: 587–95. https://doi.org/10.1016/j.chemosphere.2019.05.094.

Zheng, Jian, Chun Wang, and Fengqiang Liu. 2018. "Stereotactic brachytherapy with iodine-125 seeds plus temozolomide induced complete and durable remission in a patient with recurrent primary central nervous system lymphoma". *World Neurosurgery*, No. 117: 316–20. https://doi.org/10.1016/j.wneu.2018.06.150.

Zheng, Wei, Yan-Ming Xu, Dan-Dan Wu, Yue Yao, Zhan-Ling Liang, Heng Wee Tan, and Andy T. Y. Lau. 2018. "Acute and chronic cadmium telluride quantum dots-exposed human bronchial epithelial cells: The effects of particle sizes on their cytotoxicity and carcinogenicity". *Biochemical and Biophysical Research Communications*, No. 1: 899–903. https://doi.org/10.1016/j.bbrc.2017.11.074.

3 Mechanisms of phytoremediation and microbial remediation of radionuclides

Introduction

Biosorption of radionuclides was observed in a wide range of microorganisms. For example, uranium forms complexes with functional groups (the amino, carboxylate, phosphate, and hydroxyl) present on the microbial cell surfaces and by binding to anionic sites or precipitating as dense deposits inside the microbial cells. Besides, biosorption of plutonium (VI) was studied with bacterial biomass. It leads to a change in the oxidation state (Pu(IV)) and is bound to the phosphate groups on the cell surface.

Complexation by microorganisms was observed in the case of various radionuclides. For example, some bacteria can produce exometabolites that form soluble complexes with uranium.

The transformation of radionuclides by microorganisms is largely recognized. For example, in the case of some bacteria, reduction of pertechnetate ion to an insoluble form can occur. Furthermore, microorganisms are known to transform iodine through oxidation, volatilization, and reduction.

Many plant species have also been studied for their phytoaccumulation potential, as they can interact with many pollutants, including radionuclides, through specific mechanisms. Thus, for instance, phytoextraction is a mechanism involving the absorption of radioactive elements such as plutonium through the root system.

A description of some studies dealing with the biotreatment of three radionuclides (plutonium, uranium, and technetium), illustrating these mechanisms, will be presented in this chapter.

Plutonium

Because of the multiple oxidation states of Pu (+3, +4, +5, + 6, and +7), whose trivalent (Pu(III)) and pentavalent (Pu(V)) species are particularly soluble in water, its potential for dissolution in the environment makes it

DOI: 10.4324/9781003282600-4

of great concern. The primary source of environmental contamination by Pu is the nuclear weapons testing in the last century throughout the world. In light of this, many studies have focused on reducing the dissolution of this contaminant by using indigenous living organisms, including plants and microorganisms. The solubility, bioavailability, and mobility of Pu may be affected by direct enzymatic or indirect non-enzymatic reactions. In a study conducted by Francis and Dodge (2015, 278–83), the microbial dissolution of the polymeric form of $^{239/240}$Pu and its complex forms with Fe$^-$ and Mn$^-$ oxides and aluminosilicates present in contaminated soil from the Nevada test site was investigated in vitro conditions in the presence of citric acid and glucose at 0.5%. This work suggested that indigenous microbial activity under either aerobic or anaerobic conditions caused precipitation of Pu in solution in parallel to citric acid metabolic consumption, while in glucose-amended soil samples, a significant increase in the soluble Pu concentration was recorded. This is due to the release of Pu from complexes, such as those involving calcium carbonate or Fe$^-$ or Mn$^-$ oxides, via direct enzymatic reduction or indirectly by organic acids produced under anaerobic conditions. Another study has investigated the potential of using vetiver plant (*Vetiveria zizanoides* L. Nash) to treat soil and solution spiked with ^{239}Pu (initial activity of 100 Bq/ml) via phytoremediation and examined the ability of two chelating agents (citric acid and diethylenetriaminepentaacetic acid) to support and stimulate the remediation of the actinide under study (Singh, Fulzele, and Kaushik 2016, 141–3). Results showed that Pu removal from hydroponic solution (66.2%) was more significant than soil (7%) after 30 days of exposure. Citric acid and diethylenetriaminepentaacetic acid addition resulted in a higher per cent of Pu removal (15% and 28%, respectively) from the soil by the plant species. Besides, it was observed that the highest Pu accumulation was in the roots in both hydroponics and soil systems.

The change in the redox state of radionuclides through biotransformation can increase/decrease their toxicity by increasing/decreasing their bioavailability and their mobility (solubility). It is well established that microbes can reduce radionuclides in aerobic or anaerobic conditions, producing less soluble forms that easily precipitate. However, in some cases, the bioreduction can result in more soluble states that are environmentally threatening because of their high mobility. This has been observed in a study of the effect of different microbial species (bacteria and fungi) isolated from Pu-contaminated materials (wood and cardboard) on the oxidation states of Pu. It was found that two bacterial species – namely, *Bacillus mycoides* strain DPKI-01 and *Serratia marcescens* strain DPKI-06 – were effectively capable of mediating the aerobic reduction at low pH ($\approx$3) of 9.2×10^{-12} M ^{239}Pu (IV) to its more soluble state (Pu (III)). It was also observed that

Pu (III) and Pu (IV) were adsorbed on *Serratia marcescens* cell walls and that the two filamentous fungi species – *Absidia spinosa* var. *spinosa* and *Paecilomyces lilacinus* – were capable of adsorbing Pu (IV) onto their vegetative mycelium and spores (Lukšienė, Druteikienė, Pečiulytė, Baltrūnas, et al. 2012, 444–7). Lee, Hossner, Attrep, and Kung (2002, 174–7) compared the uptake of three [239]Pu complexes (Pu-nitrate, Pu-citrate, and Pu-diethylenetriaminepentaacetic acid (DTPA)) from three different soils using two plants (*Brassica juncea* and *Helianthus annuus*). Nitrate- and citrate-based Pu complexes and soil type were found to influence the uptake of Pu by both plants. DTPA addition improved Pu uptake by the two plants and acted in a concentration-dependent manner. In addition, of the two plants tested, *Brassica juncea* was the most efficient in Pu uptaking from the soils tested. The authors suggested that DTPA acted by chelating insoluble Pu-ions and delivering them to the plants' roots, where they are absorbed. In a study testing the potential of the lichen *Parmotrema tinctorum* of uptaking 4×10^{-4} M [239]Pu(VI), a Pu uptake of 0.040 g/g_{dry} was obtained after 96 h incubation. During the accumulation of Pu on both the upper and lower surfaces of the lichen tested, Pu(VI) was reduced to Pu (V) and then to Pu (IV) by the organic exudates released by *P. tinctorum* (Ohnuki, Aoyagi, Kitatsuji, Samadfam, et al. 2004, 344–7).

Under certain conditions, microbial-mediated bioreduction can convert the Pu from its tetravalent to its trivalent form, which is much more soluble and mobile, and thus more threatening to human health than the oxidized form. In this respect, in a study where PuPahokee peat humic acids were tested in association with a dissimilatory metal-reducing bacterium (*Shewanella putrefaciens*) to anaerobically reduce 5×10^{-9} M polymeric [239]Pu(IV) at circumneutral pH, the humic concentration of 15 mg/L was the most effective in stimulating the bioreduction of Pu(IV) to Pu(III) (102 fold higher than in control (absence of humic acids)) and acted by mediating the electron transfer from the final electron acceptor of the bacterial respiratory chain to Pu(IV) (Xie, Han, Wang, Zhou, et al. 2017, 348–50). These findings highlight the necessity to consider the negative effect of humic acids in enhancing the soluble form of Pu (Pu (III)) during anaerobic bioreduction.

The distribution coefficient factor (K_d) was used to investigate the interaction between [240]Pu(NO_3)_4 and two bacterial cultures, namely, *Desulfovibrio desulfuricans* (ATCC 7757), and a mixed culture of anaerobic bacteria at 25°C. The results showed that the mixed culture used was very effective in accumulating Pu ($K_d \approx 10^2$) at a pH ranging from 5 to 9. In addition, the authors suggested that the anionic functional groups of bacteria would be responsible for adsorbing Pu ions by ion exchange (Sasaki, Zheng, Asano, and Kudo 2001, 223–5).

Uranium

Among the four different oxidation states of uranium – U(II), U(IV), U(V), and U(VI) – U(IV) is known to be the least soluble and mobile form and can precipitate in the form of uraninite (UO_2) under reducing conditions. The interaction between bacteria and U can occur through four main mechanisms: bioreduction, biomineralization, bioaccumulation, and biosorption (Figure 3.1). All these mechanisms are potentially suitable to ensure long-term uranium removal.

Martins, Faleiro, Chaves, Tenreiro, et al. (2010, 1066–71) investigated the bio-uptake of U(VI) as uranyl acetate ($UO_2(CH_3COO)_2$) under anaerobic conditions using sulphate-reducing bacteria, microbial communities obtained from sludge and soils contaminated or not by U. Results showed that 91% of U (22 mg/L) was removed from solution by using bacterial consortium from non-contaminated soil. The microbial communities analysis revealed that bacteria capable of removing U(VI) belong to the *Rhodocyclaceae* family and *Clostridium* genus. The authors suggested that

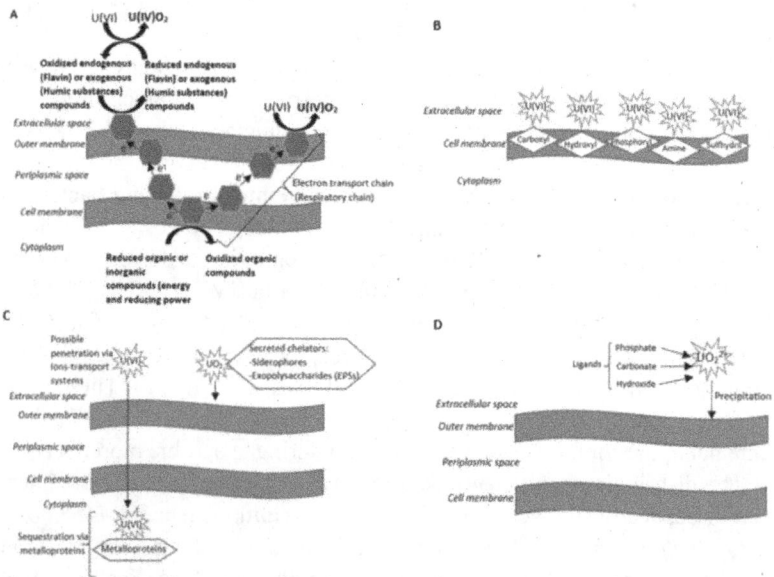

Figure 3.1 Schematic illustration of the mechanisms of the interactions between microbes and uranium. (a) Bioreduction (e.g. *Shewanella*, *Geobacter*), (b) biosorption (e.g. *Bacillus sphaericus*), (c) bioaccumulation (e.g. *Escherichia coli*, *Shewanella oneidensis*), (d) biomineralization (e.g. *Serratia*, *Pseudomonas*)

the removal of U was achieved either through precipitation, which could be the result of the bioreduction of U(VI) to U(IV), or via biosorption mediated by the functional groups (phosphoryl and/or sulphydryl groups) on the cell wall surface.

A bacterial strain isolated from a nuclear waste–contaminated site in the vicinity of a nuclear weapon test site in Gansu Province of China was investigated for its ability to accumulate U(VI). The strain identified as pertaining to *Pantoea* genus (strain TW18) exhibited a high U accumulation capacity of 79.87 mg/g at pH 4.1 and 37°C. Further, the study results clearly showed its capability in adsorbing the radionuclide via different cell surface functional groups, including carboxyl, amide, and phosphoryl groups (Zhang, Liu, Song, Ma, et al. 2018, 220–4).

Some compounds such as anthraquinone-2-sulphonate and anthraquinone-2,6-disulphonate have been described to act as extracellular electron shuttles during the microbial reduction of Fe (III) and U. The reduced form of anthraquinone-2,6-disulphonate mediator, in particular, which is generated by microorganisms via their respiratory chain, serves to transfer electrons to metals (Marshall, Matthew, Alexander, Beliaev, et al. 2010, 167–8). A study by Liu, Xie, Wang, Liu, et al. (2015, 4146–7) recorded a very high U removal rate of 99% after only 96 h at 30°C at neutral pH and under anaerobic conditions using the strain *Shewanella oneidensis* MR-1. So, the strain was capable of bioreducing U(VI) (20 mg/L) primarily to UO_2, in the presence of a quinone compound (1 mmole/L anthraquinone-2-sulphonate (AQS)) used as an extracellular electron-shuttling mediator. However, the authors reported that at concentrations exceeding 2 mmole/L AQS, the bioreduction of U was significantly inhibited, which, according to the authors, is due to an electronic competition between AQS and U(VI), which occurred at high AQS concentration. Besides, it has been found that metallic ions, such as Cu^{2-}, Cr^{6}, Mn^{2+}, hampered the U(VI) reduction as well, while Zerovalent iron (ZVI) exerted a positive effect in a dose-dependent manner. In a recent study (Tan, Li, Guo, Wang, et al. 2020, 2–4), it has been demonstrated that the strain *Leifsonia* sp. JCM28673 was capable of immobilizing uranium in the soil through reduction. The tested strain was indigenous in uranium tailings collected near Hengyang (China), from where it was isolated. The experiments were carried out in a bed-packed reactor (ø100 mm × 1000 mm) packed at a time, with glass beads (Φ 6 mm), quartz sand (Φ1–2 mm), and uranium-contaminated soil, and inoculated with the strain. It was found that *Leifsonia* sp. was able to immobilize uranium and hinder its diffusion in soil significantly.

Siderophores are biogenic iron-chelating molecules known to transport Fe(III) oxides through cell membranes (Boukhalfa, Reilly, and Neu 2007, 1019). They are also known to interact with others metals, including actinides such as uranium. Thus, hydroxamate-type siderophores secreted at their highest amount (58 µg/mg dry weight) by marine cyanobacterium *Synechococcus elongatus* BDU 130911 have been found to form complexes with uranium (VI) (as uranium acetate (1mM)), resulting in uranium sequestration in the cells (Rashmi, ShylajaNaciyar, Rajalakshmi, D'Souza, et al. 2013, 206).

In a very recent study, the pathogenic *Staphylococcus aureus* V329 biofilm-forming strain was used to bioremediate uranium. The experiments were performed in batch mode and permitted the uptake from the water of up to 47% of U(VI) (10 ppm) through acid phosphatase activity (Shukla, Hariharan, and Rao 2020, 4–5). Additionally, inorganic phosphate (in the form of potassium dihydrogen phosphate) was conductive to improving uranium removal by the pathogenic strain compared to the control. It was suggested that the inorganic phosphate generated by acid phosphatase acted by forming an insoluble stable monodentate complex with uranyl (UO_2^{2+}), which can easily precipitate.

The removal of U can be achieved with more than one mechanism in the same microorganism. This has hence been observed in a study conducted by Kazy, D'Souza, and Sar (2009, 67–9), who investigated the mechanism of U(VI) ($UO_2(NO_3)_2 \cdot 6H_2O$) removal by a strain of *Pseudomonas* (strain MTCC 3087) isolated from garden soil. The authors pointed out that the bacterial strain tested was able to sequestrate U from aqueous solution via the combination of three mechanisms: (i) biosorption through ion-exchange (Ca^{2+}, K^+/U), (ii) complexation ($U_3(PO_4)$, $U(UO_2)_3(PO_4)_2(OH)_6 \cdot 4H_2O$), and (iii) microprecipitation of crystalline U/phosphate, which functioned independently from each other.

At present, there is increasing interest in investigating the phytoremediation of uranium-contaminated media using different green plant species. Here will be described some recent studies on phytoremediation strategies for soil or waters contaminated with uranium. Also, insight will be given to those strategies combining phytoremediation with other methods (soil amendment and bioremediation). An attempt has been recently performed by Hu, Lang, Ding, Hu, et al. (2019, 60–3) to enhance the phytoremediation of uranium-contaminated soil using *Macleaya cordata* (plume poppy) in combination with three chelates (citric acid, oxalic acid, and ethylenediamine disuccinic acid). The soil used was initially artificially contaminated with 18 mg uranyl nitrate/kg soil before serving for the growth of *Macleaya cordata* seedlings. The three organic chelates were tested at 1M, and their repeated application promoted the mobility of uranium and, ultimately, its accumulation and translocation in the plant. Out of the three chelates, citric acid exhibited the highest enhancement effect on the uranium removal and led to much more accumulation of the radionuclide in shoots than in roots, with a maximum content of about 386 mg/kg.

In another study, arbuscular mycorrhiza fungi (*Glomus etunicatum* (GE, BGC NM03F)) and rhizobium bacterium (*Azorhizobium caulinodans* ORS571) have been tested for their potential to enhance the clean-up of soil artificially contaminated by uranium (VI) as uranyl-acetate by using *Sesbania rostrata*. High efficiency of the plant/microorganisms association for the removal of U was recorded, with 73.2% being removed from the soil after 60 days of incubation. On the other hand, inoculation with the two microorganisms promoted the expression of phytochelatin synthase gene in shoots demonstrating, as suggested by the authors, their great potential for the enhancement of uranium transfer from roots to shoots, where the actinide is sequestrated (Ren, Kong, Wang, and Xie 2019, 776–8).

Chen, Long, Wang, and Yang (2020, 2–7) have very recently reported the application of four different plant growth regulators (gibberellin A3, 6-benzylaminopurine, 24-epibrassinolide, and indole-3-acetic acid) for the enhanced phytoextraction of uranium from contaminated soils using mustard (*Brassica juncea* L.) as bioaccumulator. A uranium solution as uranyl acetate at 150 mg/kg was added into the soil by spraying. It was found that the remediation efficiency of uranium was significantly enhanced in the presence of the four growth regulators with the highest removal efficiency (330% greater in comparison with the controls) obtained with 500 mg/L indole-3-acetic acid.

Li, Wang, Luo, Liang, et al. (2019, 44–7) set up experiments to investigate the ability of aquatic macrophyte *Nymphaea tetragona* Georgi (Pygmy water-lily) for phytoremediation of uranium in aquatic medium. The experiments were performed in a hydroponic medium in which *Nymphaea tetragona* plantlets were cultivated and to which uranium as uranyl acetate at three concentrations (5, 30, and 55 mg/L) was added. The authors reported that plant leaves took up the highest amount of uranium (3446 mg/kg dry weight) and exhibited the highest solution-to-plant transfer factor (bioconcentration factor) (73) as well at the initial uranium concentration of 55 mg/L. However, uranium negatively affected some physiological parameters in the leaves by stimulating the activity of stress enzymes (peroxidase, superoxide dismutase, catalase, and malondialdehyde) and inhibiting the contents of soluble proteins, chlorophyll a, chlorophyll b, and carotene. However, *Nymphaea tetragona* may be a promising plant beneficial for the enhanced phytoremediation of uranium in aquatic environments despite all this.

Technetium

Although the technetium (Tc) present on the Earth is predominantly a result of anthropogenic activities (neutron-induced fission of uranium in nuclear reactors and nuclear weapons), very small quantities of natural ß-emitting

^{99}Tc has been detected as a product of spontaneous fission of ^{235}U in geologic matrices, especially in uranium ores (~10^{-12} g/g of uranium) (Dixon, Curtis, Musgrave, Roensch, et al. 1997, 1696–7; Curtis, Fabryka-Martin, Dixon, and Cramer 1999, 276). Among all the known isotopes of technetium, none of them is stable, but they are all radioactive, and the most commonly used ones are ^{99}Tc and metastable technetium (m^{99}Tc) with half-lives of 2.11 × 10^5 years and 6 hours, respectively (EPA 2002; Hidaka 2005, 249; O'Loughlin, Boyanov, Antonopoulos, and Kemner 2011, 478). In the natural environment, ^{99}Tc can occur in eight oxidation states – namely, −1, 0, +1, +3, +4, +5, +6, and +7 (O'Loughlin, Boyanov, Antonopoulos, and Kemner 2011, 479). Among these valence states, Tc(IV) and Tc(VII) are the two most critical forms in a soil system (Koch-Steindl and Pröhl 2001, 95). Tc(VII) is the most soluble form and thus is the primary-risk driving form because of its high diffusion in aquatic environments. In the environment, the mobility of Tc can be affected by various factors like a high concentration of organic matter, which acts by reducing its solubility (EPA 2002), aerobic or slightly sub-aerobic environments, in which there is the formation of a highly soluble form of Tc (pertechnetate (Tc(VII)O$_4^-$)) with high environmental mobility (O'Loughlin, Boyanov, Antonopoulos, and Kemner 2011, 479–80). On the other hand, reducing conditions (with E_{h7} ranging from +200 to +100 mV), anoxic conditions, and the presence of solid-phase associated Fe(II) seem to favour the formation of the less soluble reduced form of Tc (Tc(IV)O$_2$), which remains stable under such conditions (Icenhower, Martin, Qafoku, and Zachara 2008, 7–1; Marshall, Beliaev, and Fredrickson 2010, 97). The formation of Tc(IV)O$_2$ from Tc(VII)O$_4^-$ is shown in the following reaction (Lide 2010, 8–25):

$$TcO_4^- + 4H^+ + 3e^- \rightarrow TcO_2 + 2H_2O \ (E^\circ = 0.782 \text{ V}) \tag{3.1}$$

To date, several reports have proven the involvement of a diversity of organic, inorganic, and metal compounds (iron, nitrate, fumarate, and sulphate, manganese) reducing bacteria in the reduction of Tc(VII) into Tc(IV). These bacteria belong to gamma and delta proteobacteria; they include, but are not limited to, *Geobacter, Shewanella, Desulfovibrio, Deinococcus*, and *Anaeromyxobacter* genera (Fredrickson, Kostandarithes, Li, Plymale, et al. 2000, 2006; Lloyd, Sole, Van Praagh, and Lovley 2000, 3743; De Luca, De Philip, Dermoun, Rousset, et al. 2001, 4583; Payne and DiChristina 2006, 282; Marshall, Plymale, Kennedy, Shi, et al. 2008, 125). This group of bacteria acquires energy for supporting cellular processes by oxidation of different compounds (some are given here) coupled with Tc(VII) reduction in the respiratory chain. For instance, the oxidation of organic carbon (acetate) as a source of energy (electron donor) coupled to the reduction of TcO$_4^-$ is thermodynamically feasible with a Gibbs free-energy change (ΔG) of -436 kJ/mol of acetate (Burke, Boothman, Lloyd, Mortimer, et al. 2005, 4110).

The reduction of Tc(VII) may occur through direct enzymatic reduction involving a cytoplasmic hydrogenase. This biotic mechanism of reduction has been observed in several pure cultures of bacteria such as *Desulfovibrio desulfuricans* (Lloyd, Ridley, Khizniak, Lyalikova, et al. 1999, 2691), *Desulfovibrio fructosovorans* (De Luca, De Philip, Dermoun, Rousset, et al. 2001, 4584), *Shewanella oneidensis* MR-1 (Marshall, Plymale, Kennedy, Shi, et al. 2008, 126), and *Escherichia coli* (Lloyd, Cole, and Macaskie 1997, 2014). Tc(VII) may also be reduced indirectly by biogenic Fe(II) (Marshall, Dohnalkova, Kennedy, Plymale, et al. 2009, 540) or biogenic sulphide (Lloyd, Nolting, Sole, Bosecker, et al. 1998, 50) implicating a periplasmic hydrogenase. An additional mechanism of reduction of metals has been described involving extracellular electron shuttles such as anthraquinone-2,6-disulphonate. This quinone-based compound, which is known for its ability to enhance the electron transfer from bacteria to the final electron acceptor (Costa, Mota, Nascimento, and Dos Santos 2010, 107), was shown to be involved in Tc(VII) reduction (Lloyd, Sole, Van Praagh, and Lovley 2000, 3745). The two steps of indirect Tc(VII) reduction are shown in the two following stoichiometric reactions. First, Fe(III) is microbiologically reduced to Fe(II) using, for example, H_2 as an electron donor (3.2) (Caccavo, Blakemore, and Lovley 1992, 3212). Second, the resulting Fe(II) serves for the abiotic reduction of pertechnetate (Tc(VII)O_4^- to TcO_2 (3.3).

$$H_2 + 2Fe^{3+} \rightarrow 2Fe^{2+} + 2H^+ \qquad (3.2)$$
$$3Fe^{2+} + TcO_4^- + 4H^+ \rightarrow 3Fe^{3+} + TcO_2 + 2H_2O \qquad (3.3)$$

A significant number of studies have been dedicated to studying the biochemical mechanisms related to Tc(VII) reduction and the variable and non-variable factors affecting it. Marshall, Dohnalkova, Kennedy, Plymale, et al. (2009, 539–40) have tested a gram-negative bacterial strain (*Anaeromyxobacter dehalogenans* 2CP-C (ATTC BAA-259)) for its capacity to reduce Tc(VII)O_4^- (100 µM) to TcO_2 anaerobically by testing two electron donors – namely, acetate and H_2, which were tested at 10 mM and 41 mM, respectively. In the presence of H_2, the reduction of Tc(VII)O_4^- was quicker and more efficient than in the presence of acetate. It was suggested that the H_2-dependent Tc(VII) reduction involved a hydrogenase-mediated pathway. Transmission electron microscopy analysis revealed that the resulting TcO_2 nanoparticles were found accumulated both in the cell periplasm and outside the cell's outer membrane. Besides, it was shown that more than 80% of the initial Tc(VII) concentration tested was reduced abiotically (indirect reduction) by microbiologically produced Fe(II) within 4 hours. The authors suggested that the indirect Fe(II)-mediated Tc(VII) reduction pathway tends to occur in environments where the concentration of H_2 is low, while in the environments rich in H_2, Tc(VII) is reductively transformed through direct enzymatic reduction. The bacterial strain tested in this study could efficiently serve to control Tc mobility in sedimentary environments and soils in field treatment.

In a study by McBeth, Lear, Lloyd, Livens, et al. (2007, 193), *Geothrix* and *Geobacter* species present in sediments collected from the US Department of Energy Field Research Center (Oak Ridge, USA) were able to remove via Fe(II)-mediated reduction more than 98% Tc(VII)O$_4^-$ from a 0.5 µM solution within 121 days of incubation, in the presence of indigenous or added (20 mM acetate) electron donors. Noteworthy, the indigenous bacterial species reached a Tc(VII) removal rate > 99% within 121 days, and this in the presence of a competing electron acceptor (10 mM nitrate). This finding was supported by the results of Law, Geissler, Boothman, Burke, et al. (2010, 153), who studied the effect of nitrate concentration on the reduction of Tc(VII) by a sediment microcosm under anoxic conditions. The authors demonstrated that the activity of indigenous denitrifying bacteria resulted in the increase of pH, which promoted the removal of Tc(VII) through an abiotic reduction mechanism.

The dissimilatory iron-reducing bacterium *Geobacter sulfurreducens* has been reported to transform Tc(VII) to Tc(IV) through direct and indirect reduction. Tc(VII) (250 µM) removal via direct reduction (hydrogen-dependent reduction) in the presence of an added mixture of H_2 and CO_2 (80:20) was found to be slightly less efficient (60% within 25 h) than through indirect reduction (Fe(II)-dependent reduction) (75% within 30 h) when acetate (20 mM) was added as an electron donor. Additionally, it was demonstrated that the addition of 50 µM anthraquinone-2,6-disulphonate (synthetic electron shuttle compound) and 100 mM Fe(III) oxide resulted in the complete uptake of 250 µM Tc(VII) within only 2 h (Lloyd, Sole, Van Praagh, and Lovley 2000, 3747). Fredrickson, Kostandarithes, Li, Plymale, et al. (2000, 2008) observed a similar indirect mechanism of Tc(VII) reduction using a radiation-resistant *Deinococcus radiodurans* R1 strain. It was found that the tested strain was able to reduce different concentrations of Tc(VII) (5–100 µM) under anoxic conditions in the presence of added 0.1 mM anthraquinone-2,6-disulphonate, with a removal rate of 95% reached within 21 days.

Some studies have focused on the application of phytoremediation in the bioremediation of technetium contamination. For example, Willey, Tang, McEwen, and Hicks (2010, 761) investigated technetium (609 kBq [99]T/L) uptake by 116 angiosperm species or varieties under pot experiment conditions. The highest Tc removal (49%) was found for *Celosia argentea* shoots after 48 h of exposure time, while *Chenopodium quinoa* exhibited the highest concentration factor (61), which was calculated using the concentrations of ^{99}Tc in soil and the dry plant at harvest.

Hattink, Harms, and de Goeij (2003, 63) found that common duckweed (*Lemna minor L.*) has a high potential to remove ^{99}Tc as potassium pertechnetate from a solution medium; about 95 % of ^{99}Tc was removed from the solution within ten days of exposure to 6.2×10^{10} Bq/mol ^{99}Tc.

In a set of experiments conducted by Echevarria, Vong, Leclerc-Cessac, and Morel (1997, 947), the ability of ryegrass (*Lolium perenne L.*) and winter wheat (*Triticum aestivum L.*) to uptake ^{99}Tc as $NH_4TcO_4^-$ tested at different specific activity levels (0–250 kBq/kg dry soil) was evaluated. Up to 78% of ^{99}Tc was removed from the soil studied and accumulated in ryegrass roots. ^{99}Tc uptake was much higher in wheat leaves (up to 95%) than in the grain (1.1%).

Reference list

Boukhalfa, Hakim, Sean D. Reilly, and Mary P. Neu. 2007. "Complexation of Pu(IV) with the natural siderophore desferrioxamine B and the redox properties of Pu(IV)(siderophore) complexes". *Inorganic Chemistry*, No. 3: 1018–26. https://doi.org/10.1021/ic061544q.

Burke, Ian T., Christopher Boothman, Jonathon R. Lloyd, Robert J. G. Mortimer, Francis R. Livens, and Katherine Morris. 2005. "Effects of progressive anoxia on the solubility of technetium in sediments". *Environmental Science and Technology*, No. 11: 4109–16. https://doi.org/10.1021/es048124p.

Caccavo, Frank Jr., Richard P. Blakemore, and Derek R. Lovley. 1992. "A hydrogen-oxidizing, Fe(III)-reducing microorganism from the Great Bay Estuary, New Hampshire". *Applied and Environmental Microbiology*, No. 10: 3211–16.

Chen, Li, Chan Long, Dan Wang, and Jinyan Yang. 2020. "Phytoremediation of cadmium (Cd) and uranium (U) contaminated soils by *Brassica juncea* L. enhanced with exogenous application of plant growth regulators". *Chemosphere*, No. 242: 1–9. https://doi.org/10.1016/j.chemosphere.2019.125112.

Costa M. C., S. Mota, R. F. Nascimento, and A. B. Dos Santos. 2010. "Anthraquinone-2,6-Disulfonate (AQDS) as a catalyst to enhance the reductive decolourisation of the Azo Dyes Reactive Red 2 and Congo Red under anaerobic conditions". *Bioresource Technology*, No. 1: 105–10. https://doi.org/10.1016/j.biortech.2009.08.015.

Curtis, David, June Fabryka-Martin, Paul Dixon, and Jan Cramer. 1999. "Nature's uncommon elements: Plutonium and technetium". *Geochimica et Cosmochimica Acta*, No. 2: 275–85. https://doi.org/10.1016/S0016-7037(98)00282-8.

De Luca, Gilles, Pascale De Philip, Zorah Dermoun, Marc Rousset, and André Vervéglio. 2001. "Reduction of Technetium(VII) by *Desulfovibrio fructosovorans* is mediated by the nickel-iron hydrogenase". *Applied and Environmental Microbiology*, No. 10: 4583–7. https://doi.org/10.1128/AEM.67.10.4583-4587.2001.

Dixon, Paul, David B. Curtis, John Musgrave, Fred Roensch, Jeff Roach, and Don Rokop. 1997. "Analysis of naturally produced technetium and plutonium in geologic materials". *Analytical Chemistry*, No. 9: 1692–9. https://doi.org/10.1021/ac961159q.

Echevarria, G., P. C. Vong, E. Leclerc-Cessac, and J. L. Morel. 1997. "Bioavailability of Technetium-99 as affected by plant species and growth, application form, and soil incubation". *Journal of Environmental Quality*, No. 4: 947–56. https://doi.org/10.2134/jeq1997.00472425002600040004x.

Environmental Protection Agency (EPA). 2002. "EPA Facts about Technetium -99". Accessed March 21, 2020. www.nrc.gov/docs/ML1603/ML16032A152.pdf.

Francis, A. J., and C. J. Dodge. 2015. "Microbial mobilization of plutonium and other actinides from contaminated soil". *Journal of Environmental Radioactivity*, No. 150: 277–85. http://doi.org/10.1016/j.jenvrad.2015.08.019.

Fredrickson, J. K., H. M. Kostandarithes, S. W. Li, A. E. Plymale, and M. J. Daly. 2000. "Reduction of Fe(III), Cr(VI), U(VI), and Tc(VII) by *Deinococcus radiodurans* R1". *Applied and Environmental Microbiology*, No. 5: 2006–11. https://doi.org/10.1128/AEM.66.5.2006-2011.2000.

Hattink, Jasper, Arend V. Harms, and Jeroen J. M. de Goeij. 2003. "Uptake, biotransformation, and elimination of Tc in duckweed". *The Science of the Total Environment*, No. 1–3: 59–65. https://doi.org/10.1016/S0048-9697(02)00679-4.

Hidaka, Hiroshi. 2005. "Technetium in cosmo- and geochemical fields". *Journal of Nuclear and Radiochemical Sciences*, No. 3: 249–52.

Hu, Nan, Tao Lang, Dexin Ding, Jingsong Hu, Changwu Li, Hui Zhang, and Guangyue Li. 2019. "Enhancement of repeated applications of chelates on phytoremediation of uranium contaminated soil by *Macleaya cordata*". *Journal of Environmental Radioactivity*, No. 199–200: 58–65. https://doi.org/10.1016/j.jenvrad.2018.12.023.

Icenhower, J. P., W. J. Martin, N. P. Qafoku, and J.M. Zachara. 2008. *The Geochemistry of Technetium: A Summary of the Behavior of an Artificial Element in the Natural Environment*. Washington, DC: Pacific Northwest National Laboratory.

Kazy, Sufia K., S. F. D'Souza, and Pinaki Sar. 2009. "Uranium and thorium sequestration by a *Pseudomonas* sp.: Mechanism and chemical characterization". *Journal of Hazardous Materials*, No. 1: 65–72. https://doi.org/10.1016/j.jhazmat.2008.06.076.

Koch-Steindl, H., and G. Pröhl. 2001. "Considerations on the behaviour of long-lived radionuclides in the soil". *Radiation and Environmental Biophysics*, No. 40: 93–104. https://doi.org/10.1007/s004110100098.

Law, Gareth T. W., Andrea Geissler, Christopher Boothman, Ian T. Burke, Francis R. Livens, Jonathan R. Lloyd, and Katherine Morris. 2010. "Role of nitrate in conditioning aquifer sediments for technetium bioreduction". *Environmental Science and Technology*, No. 1: 150–5. https://doi.org/10.1021/es9010866.

Lee, J. H., L. R. Hossner, M. Attrep Jr., and K. S. Kung. 2002. "Comparative uptake of plutonium from soils by *Brassica juncea* and *Helianthus annuus*". *Environmental Pollution*, No. 2: 173–82. https://doi.org/10.1016/S0269-7491(02)00167-7.

Li, Chen, Maolin Wang, Xuegang Luo, Lili Liang, Xu Han, and Xiaoyan Lin. 2019. "Accumulation and effects of uranium on aquatic macrophyte *Nymphaea tetragona* Georgi: Potential application to phytoremediation and environmental monitoring". *Journal of Environmental Radioactivity*, No. 198: 43–9. https://doi. org/10.1016/j.jenvrad.2018.12.018.

Lide, David R., ed. 2010. *CRC Handbook of Chemistry and Physics*. Boca Raton: CRC Press/Taylor and Francis.

Liu, Jin-xiang, Shui-bo Xie, Yong-hua Wang, Ying-jiu Liu, Ping-li Cai, Fen Xiong, and Wen-tao Wang. 2015. "U(VI) reduction by *Shewanella oneidensis* mediated by anthraquinone-2-sulfonate". *Transactions of Nonferrous Metals Society of China*, No. 12: 4144–50. https://doi.org/10.1016/S1003-6326(15)64080-8.

Lloyd, J. R., J. A. Cole, and L. E. Macaskie. 1997. "Reduction and removal of heptavalent technetium from solution by *Escherichia coli*". *Journal of Bacteriology*, No. 6: 2014–21. https://doi.org/10.1128/jb.179.6.2014-2021.1997.

Lloyd, J. R., H. -F. Nolting, V. A. Sole, K. Bosecker, and L. E. Macaskie. 1998. "Technetium reduction and precipitation by sulphate-reducing bacteria". *Geomicrobiology Journal*, No. 1: 43–56. https://doi.org/10.1080/01490459809378062.

Lloyd, J. R., J. Ridley, T. Khizniak, N. N. Lyalikova, and L. E. Macaskie. 1999. "Reduction of technetium by *Desulfovibrio desulfuricans*: Biocatalyst characterization and use in a flowthrough bioreactor". *Applied and Environmental Microbiology*, No. 6: 2691–2696.

Lloyd, J. R., V. A. Sole, C. V. G. Van Praagh, and D. R. Lovley. 2000 "Direct and Fe(II)-mediated reduction of technetium by Fe(III)-reducing bacteria". *Applied and Environmental Microbiology*, No. 9: 3743–9. https://doi.org/10.1128/ aem.66.9.3743-3749.2000.

Lukšienė, Benedikta, Rūta Druteikienė, Dalia Pečiulytė, Dalis Baltrūnas, Vidmantas Remeikis, and Algimantas Paškevičius. 2012. "Effect of microorganisms on the plutonium oxidation states". *Applied Radiation and Isotopes*, No. 3: 442–9. https://doi.org/10.1016/j.apradiso.2011.11.016.

Marshall, Matthew J., Alexander S. Beliaev, and James K. Fredrickson. 2010. "Microbiological transformations of radionuclides in the subsurface". In *Environmental Microbiology*, edited by Ralph Mitchell and Ji-Dong Gu, 95–114. Hoboken: Wiley-Blackwell.

Marshall, Matthew J., Alice C. Dohnalkova, David W. Kennedy, Andrew E. Plymale, Sara H. Thomas, Frank E. Löffler, Robert A. Sanford, John M. Zachara, James K. Fredrickson, and Alexander S. Beliaev. 2009. "Electron donor-dependent radionuclide reduction and nanoparticle formation by *Anaeromyxobacter dehalogenans* Strain 2CP-C". *Environmental Microbiology*, No. 2: 534–43. https://doi. org/10.1111/j.1462-2920.2008.01795.x.

Marshall, Matthew J., Andrew E. Plymale, David W. Kennedy, Liang Shi, Zheming Wang, Samantha B. Reed, Alice C. Dohnalkova, Cody J. Simonson, Chongxuan Liu, Daad A. Saffarini, Margaret F. Romine, John M. Zachara, Alexander S. Beliaev, and James K. Fredrickson. 2008. "Hydrogenase- and outer membrane c-type cytochrome-facilitated reduction of technetium(VII) by *Shewanella oneidensis* MR-1". *Environmental Microbiology*, No. 1: 125–36. https://doi.org/10.1111/ j.1462-2920.2007.01438.x.

Martins, Mónica, Maria Leonor Faleiro, Sandra Chaves, Rogério Tenreiro, Erika Santos, and Maria Clara Costa. 2010. "Anaerobic bio-removal of uranium (vi) and chromium (vi): Comparison of microbial community structure". *Journal of Hazardous Materials*, No. 1–3: 1065–72. https://doi.org/10.1016/j.jhazmat. 2009.11.149.

McBeth, Joyce M., Gavin Lear, Jonathan R. Lloyd, Francis R. Livens, Katherine Morris, and Ian T. Burke. 2007. "Technetium reduction and reoxidation in aquifer sediments". *Geomicrobiology Journal*, No. 3–4: 189–97. https://doi. org/10.1080/01490450701457030.

O'Loughlin, Edward J., Maxim I. Boyanov, Dionysios A. Antonopoulos, and Kenneth M. Kemner. 2011. "Redox processes affecting the speciation of technetium, uranium, neptunium, and plutonium in aquatic and terrestrial environments". In *Aquatic Redox Chemistry*, edited by Paul G. Tratnyek, Timothy J. Grundl, and Stefan B. Haderlein, 477–517. Washington, DC: ACS Publications.

Ohnuki, Toshihiko, Hisao Aoyagi, Yoshihiro Kitatsuji, Mohammad Samadfam, Yasuhiko Kimura, and O. William Purvis. 2004. "Plutonium (VI) accumulation and reduction by lichen biomass: Correlation with U(VI)". *Journal of Environmental Radioactivity*, No. 3: 339–53. https://doi.org/10.1016/j.jenvrad.2004.03.015.

Payne, Amanda N., and Thomas J. Dichristina. 2006. "A rapid mutant screening technique for detection of technetium [Tc(VII)] reduction-deficient mutants of *Shewanella oneidensis* MR-1". *FEMS Microbiology Letters*, No. 259: 282–7. https://doi.org/10.1111/j.1574-6968.2006.00278.x.

Rashmi, Vijayaraghavan, Mohandass ShylajaNaciyar, Ramamoorthy Rajalakshmi, Stanley F. D'Souza, Dharmar Prabaharan, and Lakshmanan Uma. 2013. "Siderophore mediated uranium sequestration by marine cyanobacterium *Synechococcus elongatus* BDU 130911". *Bioresource Technology*, No. 130: 204–10. http://doi. org/10.1016/j.biortech.2012.12.016.

Ren, Cheng-Gang, Cun-Cui Kong, Shuo-Xiang Wang, and Zhi-Hong Xie. 2019. "Enhanced phytoremediation of uranium-contaminated soils by arbuscular mycorrhiza and rhizobium". *Chemosphere*, No. 217: 773–9. https://doi.org/10.1016/j. chemosphere.2018.11.085.

Sasaki, Takayuki, James Zheng, Hidekazu Asano, and Akira Kudo. 2001. "Interaction of Pu, Np and Pa with anaerobic microorganisms at geologic repositories". *Radioactivity in the Environment*, No. 1: 221–32. https://doi.org/10.1016/S1569-4860(01)80016-4.

Shukla, Sudhir K., S. Hariharan, and T. S. Rao. 2020. "Uranium bioremediation by acid phosphatase activity of *Staphylococcus aureus* biofilms: Can a foe turn a friend?". *Journal of Hazardous Materials*, No. 384: 1–7. https://doi.org/10.1016/j. jhazmat.2019.121316.

Singh, Shraddha, D. P. Fulzele, and C. P. Kaushik. 2016. "Potential of *Vetiveria zizanoides* L. nash for phytoremediation of plutonium (^{239}Pu): Chelate assisted uptake and translocation". *Ecotoxicology and Environmental Safety*, No. 132: 140–4. http://doi.org/10.1016/j.ecoenv.2016.05.006.

Tan, Wen-fa, Yuan Li, Feng Guo, Ya-chao Wang, Lei Ding, Kathryn Mumford, Junwen. Lv, Qin-wen Deng, Qi Fang, and Xiao-wen Zhang. 2020. "Effect of *Leifsonia* sp. on Retardation of uranium in natural soil and its potential mechanisms". *Journal of Environmental Radioactivity*, No. 217: 1–8. https://doi.org/10.1016/j. jenvrad.2020.106202.

Willey, N. J., S. Tang, A. McEwen, and S. Hicks. 2010. "The effects of plant traits and phylogeny on soil-to-plant transfer of ^{99}Tc". *Journal of Environmental Radioactivity*, No. 9: 757–66. https://doi.org/10.1016/j.jenvrad.2010.04.019.

Xie, Jinchuan, Xiaoyuan Han, Weixian Wang, Xiaohua Zhou, and Jianfeng Lin. 2017. "Effects of humic acid concentration reductive on the microbially-mediated solubilization of Pu(IV) polymers". *Journal of Hazardous Materials*, No. 339: 347–53. http://doi.org/10.1016/j.jhazmat.2017.06.054.

Zhang, Zexin, Haibo Liu, Wencheng Song, Wenjie Ma, Wei Hu, Tianhu Chen, and Lei Liu. 2018. "Accumulation of U(VI) on the *Pantoea* sp. TW18 isolated from radionuclide contaminated soils". *Journal of Environmental Radioactivity*, No. 192: 219–26. https://doi.org/10.1016/j.jenvrad.2018.07.002.

4 Mechanisms of phytoremediation and microbial remediation of heavy metals

Introduction

Metal uptaking processes of plants and microorganisms could vary based on microorganism/plant species and heavy metals. In addition, fertilization with nitrogen is advised in the phytoremediation of soils polluted with heavy metals because it improves the growth of plants and induces the production of various forms of proteins for the detoxication of heavy metals in plants. It also affects heavy metals adsorption, dissociation, and migration in the soil (Rodríguez-Ortíz, Valdez-Cepeda, Lara-Mireles, Rodríguez-Fuentes, et al. 2006, 106; Steliga and Kluk 2020).

Mechanisms of phytoremediation of heavy metals

The best plant to use in phytoremediation is that it is tolerant to high concentrations of heavy metals, has a developed root system, grows fast, produces high biomass, and accumulates heavy metals.

The mechanisms of phytoextraction of heavy metals can be summarized as follows:

1 Heavy metals extraction from the soil solution.
2 Roots uptake heavy metals through selectively permeable recognition.
3 Binding of heavy metals ions to the root cells.
4 Transport proteins take heavy metals ions to the aerial part via plant vascular system.

> (Luo, He, Polle, and Rennenberg 2016, 1131; Shahabaldin, Shazwin, Mohd, Farrah, et al. 2016, 588–9; Mahajan and Kaushal 2018, 3; Suman, Uhlik, Viktorova, and Macek 2018, 5; Manoj, Karthik, Kadirvelu, Arulselvi, et al. 2020, 2–3)

Continuous phytoextraction involves the accumulation of metal in the aerial part of the plant throughout its development cycle. To achieve this, plants

DOI: 10.4324/9781003282600-5

have efficient mechanisms for detoxifying the accumulated metal. For example, studies have shown that resistance to Ni in the alpine pennygrass *Thlaspi caerulescens* is critical for the hyperaccumulation of the metal (Milner and Kochian 2008, 3; Richau and Schat 2009, 253).

Hypertolerance is based on the fact that a substance is only harmful to an organism if it is part of its metabolism. Hyperaccumulators implement a sequestration strategy, making the element biologically unavailable (inactive), although present in the plant. Mechanisms of resistance of plants to heavy metals can be described as follows.

Chelation

The fixation of metal ions by specific ligands with high affinity reduces their concentration in the intracellular compartments and, consequently, their phytotoxicity. In plants, there are several types of ligands: organic acids, amino acids, and peptides.

Photosynthetic plants are rich in organic acids like malate, malonate, oxalate, tartrate, and isocitrate, bonding with metal ions. Thus, it has been shown in certain species that malic and malonic acids chelate most of the absorbed Ni. It was noted that in hyperaccumulating plants, Ni is preferably bound to citrate (Peñuela, Martínez, Araujo, Brito, et al. 2011, 2698; Agrawal, Lakshmanan, Kaushik, and Bais 2012, 477). Malic acid is also believed to be involved in the transport of Zn from the cytoplasmic compartment to the vacuole. After dissociation of the Zn–malate complex, the Zn would bind to the oxalate, thus forming the Zn–oxalate complex, which would represent the form of metal storage, and the malate would be returned to the cytoplasm for further transport (Sarret, Saumitou-Laprade, Bert, Proux, et al. 2002, 1815).

Two classes of peptides are capable of fixing heavy metals. On the one hand, there are metallothioneins whose amino acid composition is rich in cysteine. Their synthesis is induced by Cu, and the expression of genes encoding these proteins is strongly correlated with resistance to this metal (Grill, Winnacker, and Zenk 1987, 439; Grill, Löffler, Winnacker, and Zenk 1989, 6842). On the other hand, there are phytochelatins (PCs). These peptides are rich in cysteine and have a great affinity for cadmium. The discovery of PCs took place in 1986 by Grill, Winnacker, and Zenk (Grill, Winnacker, and Zenk 1986, 47). In this work, the researchers selected tomato cells tolerant to cadmium after successive cultures on media containing increasing concentrations of this element. These cells synthesize more PCs than non-tolerant cells. Likewise, it has been observed that the inhibition of PC synthesis by chemical processes transforms these tolerant cells into sensitive cells. The role of PCs in this tolerance to heavy metals has sparked many debates. For some authors, tolerance results from an over-production of PCs or a greater synthesis of PCs with long peptide chains. For others, it is the incorporation of sulphur (S^{2-}) into the Cd–PC complex, which makes them more stable and capable of fixing a more significant amount of cadmium. These peptides are essential for the detoxification of Cd in the mouse-ear cress *Arabidopsis thaliana*. The precipitation of Zn in the form of Zn-phytate and that of Pb by carbonates, sulphates, and phosphates has been proposed as a mechanism for the detoxification of these metals (Van Steveninck, Van Steveninck, Fernando, Horst, et al. 1987, 247; Keunen, Truyens, Bruckers, Remans, et al. 2011, 1084; Bruno, Pacenza, Forgione, Lamerton, et al. 2017, 2; Fischer, Spielau, and Clemens 2017, 8; Yao, Cai, Yu, and Liang 2018, 13).

Certain metals, such as cadmium, cause the biosynthesis of ethylene in the roots and leaves. Ethylene would then be a messenger stimulating lignification capable of limiting the flow of metals in the vascular systems and accelerating the antioxidant response by induction of ascorbate peroxidase activity and the synthesis of metallothionein (di Toppi and Gabbrielli 1999, 112).

Heat shock proteins (Hsp), already known for their involvement in stress linked to heat shock, are also responsible for tolerance to toxic metals. Indeed, a wide variety of stresses, having in common the denaturing of proteins (proteotoxic stress), can induce this response of the "thermal shock" type. The induction of Hsp by proteotoxic stress allows the cell to repair the protein damage caused by resolubilization of the aggregates, renaturation of the polypeptides, or, if this is impossible, the engagement of the denatured proteins towards the degradation pathways. Thus, the results obtained on tomato (*Lycopersicon esculentum*) cells have established that the action of H_2O_2 induces the synthesis of Hsp (Banzet, Richaud, Deveaux, Kazmaier, et al. 1998, 519). Similarly, in the tomato (*Lycopersicon peruvianum*) cells exposed to 1 mM cadmium, significant amounts of hsp70 were found in the plasmalemma, the mitochondrial membrane, and the endoplasmic reticulum, sites of multiple damages from oxidative stress caused by heavy metals (Neumann, Lichtenberger, Gunther, Tschiersch, et al. 1994, 360).

Compartmentalization

Various studies have shown the importance of forming the PCs–Cd–S^{2-} complex in the mechanisms of tolerance to Cd. The presence of the two types of complexes called LMW (Low Molecular Weight) or PCs–Cd and HMW (High Molecular Weight) or PCs–Cd–S^{2-} were first observed in yeast (Reese and Wagner 1987, 241; Reese and Winge 1988, 12832; Reese, White, and Winge 1992, 225).

Work carried out on tobacco plants has shown that the Cd–PC complex and all of the Cd present in the protoplast are localized in the vacuole. Thus, for these authors, PCs do not play the simple role of chelator in the cytoplasmic medium but are probably involved in the metal transport towards the vacuole. In this context, a transporter using ATP has been located at the tonoplast level and is responsible for transporting the PCs and the PCs–Cd complexes inside the vacuole. In the vacuole, these compounds will incorporate Cd^{2+} and S^{2-} to form the PCS–Cd–S^{2-} complex of higher molecular weight (Zanella, Fattorini, Brunetti, Roccotiello, et al. 2016, 605).

Biotransformation

The chemical transformation of certain elements (Se, (Cr(VI)), and As) and their incorporation into the metabolic pathways make these elements very toxic to plants (Zwolak 2020, 44). The toxicity of Se is due to its transformation into selenocystein and selenomethionin, which will replace cysteine and methionine in protein synthesis (Plateau, Saveanu, Lestini, Dauplais, et al. 2017, 1). However, certain species of *Astragalus* are capable of reducing Se and can thus accumulate it in a non-toxic form in the aerial part (Broyer, Johnson, and Huston 1972, 635). Selenocysteine methyltransferase, an enzyme responsible for the methylation of selenocystein, has been isolated and characterized in these plants (Sors, Martin, and Salt 2009, 110).

Bioavailability, absorption, and translocation

The resistance of the plants, although essential for phytoremediation, is not sufficient to ensure a substantial accumulation of metals in the aerial part of the plant. The bioavailability of these elements, their absorption by the roots, and their translocation are also essential for improving this practice. The concentration of metal ions in the soil solution depends mainly on the pH, organic or inorganic matter content, and redox potential. Thus, it is recognized that high pH promotes the adsorption and association processes of metallic elements, such as Cd, Zn, Pb, Fe, Mn, and Cu, to organic matter or other constituents of the solid phase of the ground. The addition of chelators to increase the bioavailability of metal ions is essential for phytoextraction. In addition, acid root excretion can lower the pH and consequently increase the bioavailability of metals in the rhizosphere.

Plants have developed several strategies to increase the bioavailability of essential trace elements. This is illustrated by the mechanisms used to acquire iron and other mineral elements. These strategies include producing metal chelators (phytosiderophores) such as mugineic acid or avenic acid produced by the plant in response to iron deficiency and probably also in

Zinc. At the level of the rhizosphere, phytosiderophores fix and mobilize Fe, Cu, Zn, and Mn. The passage through the cell membrane takes the form of a metal–phytosiderophore complex using specialized transporters (Schaaf, Erenoglu, and von Wirén 2004, 989; Meda, Scheuermann, Prechsl, Erenoglu, et al. 2007, 1761).

Simultaneous active and passive transport of Cd has been proposed for soybean, lupin, and corn roots. Recent work has shown that passive transport could correspond to the displacement of Cd by the apoplasmic pathway and that it would be negligible at low concentrations. The other type of transport uses a protein-based transporter that would exist at the plasma membrane (Page, Weisskopf, and Feller 2006, 339; Ling, Gao, Du, Zhao, et al. 2017, 220; Xu, Liu, He, Xu, et al. 2017, 1).

Once absorbed by the roots, metal ions can accumulate there or be transported to the aerial part. The upward transport of metals takes place mainly through the xylem. The level of cadmium in organs, other than the roots, depends on the amount absorbed from this element, the intracellular distribution of the metal, and its translocation from the roots to the aerial parts. In some plants, such as tomatoes, peppers, corn, and barley, Cd is preferentially accumulated in the roots. However, in some plants, cadmium can accumulate strongly outside the roots; this is the case of tobacco, used in the manufacture of cigarettes, where we can find up to 80% of the Cd accumulated in the leaves. The transport of cadmium in the xylem sap would be dependent on the transpiration current. Data suggest that Cd is transported in the form of a Cd–citrate complex (Gichner, Patková, Száková, and Demnerová 2004, 56; Xie, Hu, Du, Sun, et al. 2014, 2–3; Song, Jin, and Wang 2016, 133; Yang, Ge, Zeng, Zhou, et al. 2017, 2; Sidhu, Bali, and Bhardwaj 2019, 269).

The isolation of a Ni–citrate complex from a hyper-accumulating plant reinforces the role of organic acids in the transport of metals. Citrate is also implicated in the translocation towards PS (PhytoSiderophores) of Iron and Zn^{2+}, whereas for Cu^{2+}, it is amino acids such as histidine or asparagine which play the role of the chelator. Other chelators could also play an essential role in the mobility of the metal ion inside the plant. This is the case with the nicotianamine acid present in all plants and which can transport divalent cations such as Cu^{2+}, Ni^{2+}, Co^{2+}, Zn^{2+}, Fe^{2+}, and Mn^{2+} (Reichman and Parker 2005, 129–30).

Some authors have noted an accumulation of metals in old leaves before their abscission, decreasing the plant's metal concentrations. In general, the concentration in the leaves increases with age. This is the case of *Armeria maritima* ssp. *halleri*; in brown (older) leaves, the concentrations of Cu, Cd, Zn, and Pb are three to eight times those of young leaves. This observation suggests an internal transport from green leaves, still active from a photosynthetic point of view, towards the leaves that are about to fall, preserving photosynthesis while detoxifying the plant (Dahmani-Muller, van Oort, Gélie, and Balabane 2000, 231). An accumulation of Cd has also been observed in the trichomes of several plants (Salt, Prince, Pickering, and Raskin 1995, 1427).

Phytostabilization

Phytostabilization consists of installing plants tolerant of the presence of heavy metals in the soil. They also limit erosion and prevent dust from entering the atmosphere. They can also secrete substances (various kinds of sugars, polysaccharides, organic and amino acids, peptides, and proteins depending upon the plants) that chemically stabilize heavy metals in the rhizosphere, particularly preventing their migration runoff and groundwater. Plants that can also accumulate heavy metals in their root system are interesting for phytostabilization (Alkorta, Becerril, and Garbisu 2010, 138; Shackira and Puthur 2019, 264). A field cultivation experiment was conducted to explore the potential of seagrass (*Zostera marina*) transplants to remediate two bay sediments contaminated by Cu, As, Pb, Fe, Cd, Co, Zn, and Hg. The results indicated the ability of the tested transplants to hyperaccumulate Zn, Co, Cu, Mn, and Cd, as evidenced from their high concentrations in leaves (Lee, Suonan, Kim, Hwang, et al. 2019, 10). In a very recent study, Aztec Marigold (*Tagetes erecta* L.) was examined for its ability to uptake three heavy metals – namely, Cd, Pb and Zn – from lateritic soil in India. The results showed that the plant is a shoot hyperaccumulator of Cd and Zn with a bioconcentration factor > 1 for both metals and a translocation factor > 1 for Cd (Madanan, Shah, Varghese, and Kaushal 2021, 19).

Phytovolatilization

Plants naturally can absorb contaminants. Then, during metabolism, contaminants or their derivatives are associated with volatile compounds, which are ultimately released into the atmosphere. An example is the treatment of sites contaminated with Se. The volatilization of Se from plant tissues is a mechanism for the detoxification of this element. The transformation of Se into nontoxic volatile compounds has been reported in plants. Se is volatilized by hyperaccumulative plants in the form of dimethyl diselenide or dimethyl selenide (Evans, Asher, and Johnson 1968, 13; Limmer and Burken 2016, 6633).

Mechanisms of bioremediation of heavy metals by microorganisms

The ability to cope with the toxicity of heavy metals is due to the intrinsic properties of the microorganisms. Resistance is the ability to survive heavy metals by detoxification mechanisms in direct response to the presence of metals in the environment (Gadd 1992, 200).

Reactivity of bacteria towards metals

Microorganisms, especially bacteria, can interact with metals through different mechanisms. For example, we can observe a transformation of metals by oxidation/reduction or alkylation processes. These modifications generally modify the toxicity and mobility of the original metal. Metals can also be accumulated by passive adsorption phenomena (independent of metabolism) or active transport inside the cell (dependent on metabolism). The production by microorganisms of substances such as organic compounds or sulphides, which modify the solubility and therefore the mobility of metals, has also been observed. In addition, through their participation in biogeochemical cycles, microorganisms modify the characteristics of organic matter in their environment, which can modify the behaviour of metals via chelation or complexation mechanisms. Moreover, bacteria can indirectly influence the mobility of metals by modifications (by acidification, for example) of the environment (Chang, Law, and Chang 1997, 1651; Ledin 2000, 1–3).

The interactions between bacterial cells and metals are governed by passive or active mechanisms (Haferburg and Kothe 2007, 456). The former are independent of metabolism and, therefore, of the physiological state of cells (living or dead); they are rapid and reversible. They occur at the cell/solution interface and involve ion exchange, surface complexation, or precipitation. They depend on the metabolism of the cells and are therefore specific to each bacterial strain; they are slower and generally inducible. These passive and active interactions will depend on the cellular structure and can coincide. In general, it is considered that heavy metals can be fixed in the cellular structure and consequently biosorbed on binding sites. Biosorption or "passive uptake" is independent of metabolism. Also, heavy metals can enter cells by passing through the membrane through metabolism. This mode of assimilation is known as assimilation or "active uptake". These two modes of interaction are more generally grouped under bioaccumulation (Malik 2004, 262; Ali, Khan, and Sajad 2013, 869).

In metabolically active cells, assimilation takes place in two phases. First, an initial phase of rapid biosorption, followed by a slower phase of active assimilation depending on the metabolism and the metals considered. Unfortunately, most studies on the biosorption of metals on bacterial cells have been carried out without considering the effect of the physiological activity of the cells (Volesky and Holan 1995, 237; Kratochvil and Bohumil 1998, 299; Ismail and Moustafa 2016, 131).

Passive uptake

The negative net charge of the cell envelope makes some microorganisms capable of fixing and accumulating metal cations in the environment. The wall of bacterial cells has many functional groups capable of reacting with the external constituents. Indeed, being in contact with the external environment, the various components present on the outer layer of cells are places of privileged interactions with the environment (ions, metals, pesticides, soil components, etc.). The functional groups having anionic functions, at the level of the bacterial surface, are the main ones responsible for the ability of bacteria to fix metals (Fein, Daughney, Yee, and Davis 1997, 3326; Yee and Fein 2001, 2038; Guiné, Spadini, Sarret, Muris, et al. 2006, 1812; Guiné, Martins, Causse, Durand, et al. 2007, 267; Mishra, Boyanov, Bunker, Kelly, et al. 2009, 4311).

The following are the main reactive sites:

• Lipopolysaccharides (LPS) are chains of molecules evolving outside of cells. They are composed of a common lipid A on which is fixed a polysaccharide, which contains carboxyl groups and phosphomonoesters.
• Phospholipids are composed of hydrophobic chains attached to a glycerol unit on which is attached a hydrophilic group. On the latter, there are two reactive sites (amine and phosphodiester functions).
• The peptidoglycan (PG) network is a rigid structure composed of two sugar molecules (N-acetylglucosamine and N-acetylmuramic acid) linked to a short peptide (tetrapeptide to octapeptide). It contains three carboxyl functions and an amine

(Cox, Smith, Warren, and Ferris 1999, 4514;
Yee and Fein 2001, 2037).

It should be noted that the constitutive wall differences between Gram-positive and Gram-negative cells seem to have a minor influence on the sorption capacities of metals (Yee and Fein 2001, 2037; Ngwenya, Sutherland, and Kennedy 2003, 537).

In 1997, Fein et al. (Fein, Daughney, Yee, and Davis 1997, 3326) proposed a universal complexation model for metals with three surface sites: carboxyls, phosphates, and hydroxyls or amines. This model was supported by subsequent works (Kelly, Boyanov, Bunker, Fein, et al. 2001, 946; Ngwenya, Sutherland, and Kennedy 2003, 527; Guiné, Spadini, Sarret, Muris, et al. 2006, 1806; Ueshima, Ginn, Haack, Szymanowski, et al. 2008, 5885; Mishra, Boyanov, Bunker, Kelly, et al. 2009, 4311).

Active uptake

Bacteria actively take up essential metals in ionic form to meet a specific physiological need (metabolism, stress response). This sampling is done in two stages: the metal ions are initially adsorbed on the extracellular surface (a passive mechanism), as previously explained, and then specialized transport systems (transport proteins), which are generally accompanied by an energy expenditure (ATP hydrolysis), are activated to bring the metal inside the cell (an induced mechanism) (Nies 1999, 730). Because of the similarities between metals, non-essential metals (toxic metals have a higher affinity for nitrogen and sulphur containing ligands and form bonds of covalent character) can accidentally follow the same path as essential metals (presenting a higher affinity for oxygen containing ligands and usually form ionic bonds with the ligands), which partly explains the toxicity of metals. Internalization is the crucial step in the metal assimilation mechanism. The formation of complexes between the metal and the anionic functional groups is a prerequisite for the assimilation of metals by microorganisms. Once the metal is sorbed, it can then be transported into the periplasmic space and possibly into the cytoplasm. This involves many complexing reactions and molecules necessary for transport, storage, surface bonds, and biological functions.

There are three main families of carriers: ATPases (Adenosine TriPhosphatase), RND (Resistance-Nodulation-cell Division), and CDF (Cation Diffusion Facilitators) (Nies 1999, 731; Worms, Simon, Hassler, and Wilkinson 2006, 1725; Haferburg and Kothe 2007, 454).

Survival strategies for resistant bacteria

Metals with no biological function are generally tolerated at low concentrations, while essential metals are accepted at higher concentrations. The latter participate in the metabolic functioning of cells as constituents of enzymes or structural constituents (the membrane, for example). Thus, the concentration and speciation of the metal determine whether it is beneficial or harmful to the cell. Therefore, control of internal concentrations, or homeostasis, is necessary (Bruins, Kapil, and Oehme 2000, 198).

Bacteria have developed different defined strategies to protect themselves from the toxicity of metals:

- Many bacteria are now known for their ability to excrete metals through efflux systems. These transporters are characterized by a strong affinity for the substrate and make it possible to maintain low metal concentrations in the cytosol (Mergeay, Nies, Schlegel, Gerits, et al. 1985, 333; Nies and Silver 1995, 186; Nies 2003, 313). One of the best-known examples is the bacteria *Cupriavidus metallidurans* CH34 (renamed from *Ralstonia metallidurans*), the subject of numerous studies. Three main families of proteins responsible for efflux in microorganisms have been described:

    - Proteins of the RND family, the first discovery of which was the protein CzcCBA of the bacteria *C. metallidurans* in which the plasmid pMOL30 allows resistance to Co, Zn, and cadmium via an efflux mechanism. It is likewise for the protein allowing resistance to Co and Ni carried by the plasmid pMOL28. The effectiveness of this system is because not only does it decrease the cytoplasmic concentration but also the periplasmic concentration. In this case, the cations can be excreted before they even enter the cell (Saier, Tam, Reizer, and Reizer 1994, 841; Nies 2003, 315).

- The CDF family, the primary substrates of which are zinc, Co, Ni, cadmium, and iron. Proteins of this family are generally responsible for the efflux of metals present in the cytoplasm (Lee, Grass, Haney, Fan, et al. 2002, 273). The concentration gradient controls these efflux systems. All of the CDF proteins described in bacteria are involved in resistance to Zn. This is the case of the CzcD protein in *C. metallidurans* CH34 (Anton, Weltrowski, Haney, Franke, et al. 2004, 7499). In *Escherichia coli*, the protein ZitB has been demonstrated for its involvement in resistance to Zn (controlled by the potassium gradient and the "proton motive force"), the expression of this protein is induced by the presence of Zn.

- The third major family involved in the efflux of metals is that of "P-type ATPases", which form a large family of active transporters whose energy comes from the hydrolysis of ATP. There are ATPases that import and others that export metals. For example, the Zn-CPx-type ATPases are involved in Zn, Cd, and Pb transport mainly from the cytoplasm to the external environment or the periplasm. The first protein described was *cadA*, which is involved in the efflux of Cd, discovered in *Staphylococcus aureus* (Yoon and Silver 1991, 7636; Gaballa and Helmann 2003, 497).

- Microorganisms can produce and secrete organic (extracellular polymeric substance (EPS)) or inorganic substances (metabolites), which are likely to modify the mobility of metals either by immobilizing them (precipitation, adsorption) or by (re)mobilizing them (solubilization) (Chen, Lion, Ghiorse, and Shuler 1995, 421–2; Gilmour and Riedel 2009, 8; Uroz, Calvaruso, Turpault, and Frey-Klett 2009, 378; Gadd 2010, 609). This mechanism is often described as bioweathering or bioleaching (Burford, Fomina, and Gadd 2003, 1128; Sand and Gehrke 2006, 50).

- If toxic metals have entered the cell and cannot be excreted by efflux systems, several microorganisms have developed cytosolic sequestration mechanisms to protect themselves. It has been shown in many metal-resistant microorganisms that internal compounds, such as polyphosphate granules or thiol groups (containing sulphur), were capable of sequestering large quantities of metal cations (Keasling, Van Dien, and Pramanik 1998, 232; Finlay, Allan, Conner, Callow, et al. 1999, 87; Gadd 2000, 271; Pagès, Sanchez, Conrod, Gidrol, et al. 2007, 2829). The bioaccumulation of heavy metals and their subsequent storage in the cell in inert form allow the cell to decrease their toxicity. This is, for example, the case in *C. metallidurans* CH34, which reduces selenite in the form of elemental Se (red) and accumulates it in the form of nodules in the cytoplasm (Roux, Sarret, Pignot-Paintrand, Fontecave, et al. 2001, 769; Sarret, Avoscan, Carriere, Collins, et al. 2005, 2331).

- Extracellular precipitation occurs when microorganisms produce or secrete substances that react with soluble metals to produce an insoluble metal compound. Inorganic metabolites such as sulphate, carbonate, or phosphate ions derived mainly from respiratory metabolism can precipitate toxic metal ions. The formation of metallic sulphides by sulphato-reductive bacteria (anoxic sediments or poorly aerated soils) is, for example, one of the best-known microbiological immobilization processes (Ledin 2000, 3). In order to be effective, co-precipitation must significantly decrease the concentration of dissolved elements, that is, below the concentration for which bacterial growth is affected by the concentration of metals (Mugwar and Harbottle 2016, 243).

- Metals can be biotransformed by redox mechanisms linked to cellular respiration (this is the case for iron Fe and manganese Mn) or by alkylation (this is the case for mercury Hg). These transformations are significant for certain bacteria (sulphato-reducing bacteria, mainly) and impact the bioavailability, mobility, and toxicity of the metal (depending on its speciation). Toxic metals can also be transformed into a less toxic or even non-toxic form by oxidation or enzymatic reduction. For their energy metabolism, many prokaryotes can use the metals present in different oxidation states (As, Co, Cr, Cu, Fe, Mn, or Se) as electron donors or acceptors (Ledin 2000, 16; Martin 2019, 807).

EPS are polymers produced by prokaryotic microorganisms and eukaryotes in a natural or artificial environment. Although not essential for free cells in the natural environment, EPS play an essential role in cell adhesion and the formation of cellular aggregates (biofilms, sludges, biogranules) and protect cells from environmental aggressions (Bhaskar and Bhosle 2006, 196; Kenney and Fein 2011, 109).

Bacteria generally overproduce EPS in response to environmental stress (Guibaud, Comte, Bordas, Dupuy, et al. 2005, 636). EPS comprises a wide variety of high molecular weight macromolecules such as polysaccharides (75–90% by mass), proteins, nucleic acids and phospholipids, and some non-polymeric low molecular weight molecules. These compounds may be present in its pure state or in connection with different functional groups. Often, proteins combine with oligosaccharides to form glycoproteins or trace elements to form lipoproteins (Wingender, Neu, and Flemming 1999, 2–4). In general, polymers outside the cell wall and not directly anchored in the membrane are considered EPS. There are two types of EPS: bound EPS and soluble EPS. Soluble EPS are actively secreted by bacteria and are biodegradable, while bound EPS remain attached to the active biomass or are molecules resulting from cell lysis (Guibaud, Tixier, Bouju, and Baudu. 2003, 1701; Guibaud, Comte, Bordas, Dupuy, et al. 2005, 636). EPS have very good metal retention properties with variations in specificity and affinity. The attachment of cations to bacterial biopolymers is generally done by electrostatic interaction with negatively charged functional groups such as uronic acids, phosphoryl groups associated with membrane components, or amino acid carboxylic groups. In addition, cations can also be fixed by negatively charged polymers through hydroxyl groups. EPS of aquatic microorganisms act as polyanions under natural conditions by forming salt bridges with carboxyl groups of acid polymers (polysaccharides containing uronic acids) or forming weak electrostatic bonds with hydroxyl groups on polymers containing neutral carbohydrates. While a large number of metals are known to bind to polysaccharides, the protein part of EPS also plays a significant role in the complexation of metal ions. Proteins rich in acidic amino acids (glutamic acid and aspartic acid) also contribute to the anionic properties of EPS (Comte, Guibaud, and Baudu 2006, 815; Pal and Paul 2008, 54).

Very rich in negatively charged groups (pyruvate, phosphates, succinate, hydroxyls, and uronic acids) that bind strongly to metal ions, EPS are generally acidic, very reactive with metals, and very mobile; they can be assimilated to dissolved ligands and can thus participate in the mobilization of metals (Chen, Lion, Ghiorse, and Shuler 1995, 422).

The EPS produced by many microorganisms are of great interest in the processes of bioremediation by their participation in the flocculation and the adhesion of metal ions in solution (Bala Subramanian, Yan, Tyagi, and Surampalli 2010, 2254; Costa, Raaijmakers, and Kuramae 2018, 2; Nouha, Kumar, Balasubramanian, and Tyagi 2018, 226).

Reference list

Agrawal, Bhavana, Venkatachalam Lakshmanan, Shail Kaushik, and Harsh P. Bais. 2012. "Natural variation among *Arabidopsis* accessions reveals malic acid as a key mediator of Nickel (Ni) tolerance". *Planta*, No. 2: 477–89. 10.1007/s00425-012-1621-2.

Ali, Hazrat, Ezzat Khan, and Muhammad Anwar Sajad. 2013. "Phytoremediation of heavy metals-concepts and applications". *Chemosphere*, No. 7: 869–81. https://doi.org/10.1016/j.chemosphere.2013.01.075.

Alkorta, Itziar, José Maria Becerril, and Carlos Garbisu. 2010. "Phytostabilization of metal contaminated soils". *Reviews on environmental health*, No. 2: 135–46. 10.1515/REVEH.2010.25.2.135.

Anton, Andreas, Annett Weltrowski, Christopher J. Haney, Sylvia Franke, Gregor Grass, Christopher Rensing, and Dietrich H. Nies. 2004. "Characteristics of zinc transport by two bacterial cation diffusion facilitators from *Ralstonia metallidurans* CH34 and *Escherichia coli*". *Journal of Bacteriology*, No. 22: 7499–07. 10.1128/JB.186.22.7499-7507.2004.

Bala Subramanian, S., S. Yan, R. D. Tyagi, and R. Y. Surampalli. 2010. "Extracellular polymeric substances (EPS) producing bacterial strains of municipal wastewater sludge: Isolation, molecular identification, EPS characterization and performance for sludge settling and dewatering". *Water Research*, No. 7: 2253–66. 10.1016/j.watres.2009.12.046.

Banzet, Nathalie, Christiane Richaud, Yves Deveaux, Michael Kazmaier, Jean Gagnon, and Christian Triantaphylides. 1998. "Accumulation of small heat shock proteins, including mitochondrial HSP22, induced by oxidative stress and adaptative response in tomato cells". *Plant Journal*, No. 4: 519–27. https://doi.org/10.1046/j.1365-313X.1998.00056.x.

Bhaskar, P. V., and Narayan B. Bhosle. 2006. "Bacterial extracellular polymeric substances (EPS): A carrier of heavy metals in the marine food chains". *Environment International*, No. 32: 191–8. https://doi.org/10.1016/j.envint.2005.08.010.

Broyer, T. C., C. M. Johnson, and R. P. Huston. 1972. "Selenium and nutrition of *Astragalus*". *Plant Soil*, No. 36: 635–49. https://doi.org/10.1007/BF01373513.

Bruins, Mark R., Sanjay Kapil, and Frederick W. Oehme. 2000. "Microbial resistance to metals in the environment". *Ecotoxicology and Environmental Safety*, No. 45: 198–207. https://doi.org/10.1006/eesa.1999.1860.

Bruno, Leonardo, Marianna Pacenza, Ivano Forgione, Liam R. Lamerton, Maria Greco, Adriana Chiappetta, and Maria B. Bitonti. 2017. "In *Arabidopsis thaliana* cadmium impact on the growth of primary root by altering SCR expression and auxin-cytokinin cross-talk". *Frontiers in Plant Science*, No. 8: 1323. https://doi.org/10.3389/fpls.2017.01323.

Burford, E. P., Marina Fomina, and G. M. Gadd. 2003. "Fungal involvement in bioweathering and biotransformation of rocks and minerals". *Mineralogical Magazine*, No. 6: 1127–55. https://doi.org/10.1180/0026461036760154.

Chang, Jo-Shu, Robin Law, and Chung-Cheng Chang. 1997. "Biosorption of lead, copper and cadmium by biomass of *Pseudomonas aeruginosa* PU21". *Water Research*, No. 31: 1651–8. https://doi.org/10.1016/S0043-1354(97)00008-0.

Chen, Jyh-Herng, Leonard W. Lion, William C. Ghiorse, and Michael L. Shuler. 1995. "Mobilization of adsorbed cadmium and lead in aquifer material by bacterial extracellular polymers". *Water Research*, No. 29: 421–30. https://doi.org/10.1016/0043-1354(94)00184-9.

Comte, Sophie, Gilles Guibaud, and Michel Baudu. 2006. "Biosorption properties of extracellular polymeric substances (EPS) resulting from activated sludge according to their type: Soluble or bound". *Process Biochemistry*, No. 41: 815–23. https://doi.org/10.1016/j.procbio.2005.10.014.

Costa, Ohana Y. A., Jos M. Raaijmakers, and Eiko E. Kuramae. 2018. "Microbial extracellular polymeric substances: Ecological function and impact on soil aggregation". *Frontiers in Microbiology*, No. 9: 1636. https://doi.org/10.3389/fmicb.2018.01636.

Cox, Jenny S., D. Scott Smith, Lesley A. Warren, and F. Grant Ferris. 1999. "Characterizing heterogeneous bacterial surface functional groups using discrete affinity spectra for proton binding". *Environmental Science & Technology*, No. 33: 4514–21. https://doi.org/10.1021/es990627l.

Dahmani-muller, H., Folkert van Oort, B. Gélie, and M. Balabane. 2000. "Strategies of heavy metal uptake by three plant species growing near a metal smelter". *Environmental Pollution*, No. 2: 231–8. https://doi.org/10.1016/s0269-7491(99)00262-6.

di Toppi, Luigi Sanita, and R. Gabbrielli. 1999. "Response to cadmium in higher plants". *Environmental and Experimental Botany*, No. 2: 105–30. https://doi.org/10.1016/S0098-8472(98)00058-6.

Evans, Christine S., C. J. Asher, and C. M. Johnson. 1968. "Isolation of dimethyl diselenide and other volatile selenium compounds from *Astragalus racemosus* (Pursh.)". *Australian Journal of Biological Sciences*, vol. 21: 13–20. https://doi.org/10.1071/BI9680013.

Fein, Jeremy B., Christopher Daughney, Nathan Yee, and Thomas A. Davis. 1997. "A chemical equilibrium model for metal adsorption onto bacterial surfaces". *Geochimica et Cosmochimica Acta*, No. 61: 3319–28. https://doi.org/10.1016/S0016-7037(97)00166-X.

Finlay, John A., Victoria J. M. Allan, Alex Conner, Maureen E. Callow, Gabriela Basnakova, and Lynne E. Macaskie. 1999. "Phosphate release and heavy metal accumulation by biofilm-immobilized and chemically-coupled cells of a *Citrobacter sp.* pre-grown in continuous culture". *Biotechnology and Bioengineering*, No. 1: 87–97. https://doi.org/10.1002/(SICI)1097-0290(19990405)63:1<87::AID-BIT9>3.0.CO;2-0.

Fischer, Sina, Thomas Spielau, and Stephan Clemens. 2017. "Natural variation in *Arabidopsis thaliana* Cd responses and the detection of quantitative trait loci affecting Cd tolerance". *Scientific Reports*, No. 7: 3693. https://doi.org/10.1038/s41598-017-03540-z.

Gaballa, A., and J. D. Helmann. 2003. "*Bacillus subtilis* CPx-type ATPases: Characterization of Cd, Zn, Co and Cu efflux systems". *Biometals*, No. 16: 497–505. https://doi.org/10.1023/A:1023425321617.

Gadd, Geoffrey Michael. 1992. "Metals and microorganisms: A problem of definition". *FEMS Microbiology Letters*, No. 79: 197–203. https://doi.org/10.1111/j.1574-6968.1992.tb14040.x.

Gadd, Geoffrey Michael. 2000. "Bioremedial potential of microbial mechanisms of metalmobilization and immobilization". *Current Opinion in Biotechnology*, No. 11: 271–9. https://doi.org/10.1016/S0958-1669(00)00095-1.

Gadd, Geoffrey Michael. 2010. "Metals, minerals and microbes: Geomicrobiology and bioremediation". *Microbiology*, No. 156: 609–43. https://doi.org/10.1099/mic.0.037143-0.

Gichner, Tomás, Zdenka Patková, Jirina Száková, and Katerina Demnerová. 2004. "Cadmium induces DNA damage in tobacco roots, but no DNA damage, somatic mutations or homologous recombination in tobacco leaves". *Mutation Research*, No. 559: 49–57. https://doi.org/10.1016/j.mrgentox.2003.12.008.

Gilmour, C., and G. Riedel. 2009. "Biogeochemistry of trace metals and mettaloids". In *Encyclopedia of Inland Waters*, edited by Gene E. Likens, 7–15. Cambridge, MA: Academic Press Elsevier Inc. https://doi.org/10.1016/B978-012370626-3.00095-8.

Grill, Erwin, S. Löffler, Ernst-Ludwig Winnacker, and Meinhart H. Zenk. 1989. "Phytochelatins, the heavy-metal-binding peptides of plants, are synthesized from glutathione by a specific g-glutamylcysteine dipeptidyl transpeptidase (phytochelatin synthase)". *Proceedings of the National Academy of Sciences USA*, No. 86: 6838–42. https://doi.org/10.1073/pnas.86.18.6838.

Grill, Erwin, Ernst-Ludwig Winnacker, and Meinhart H. Zenk. 1986. "Homophytochelatins are heavy metal-binding peptides of homo-glutathione containing fabales". *FEBS Letters*, No. 205: 47–50. https://doi.org/10.1016/0014-5793(86)80863-8.

Grill, Erwin, Ernst-Ludwig Winnacker, and Meinhart H. Zenk. 1987. "Phytochelatins, a class of heavy-metal-binding peptides from plants, are functionally analogous to metallothioneins". *Proceedings of the National Academy of Sciences USA*, No. 2: 439–43. https://doi.org/10.1073/pnas.84.2.439.

Guibaud, Gilles, N. Tixier, A. Bouju, and Michel Baudu. 2003. "Relation between extracellular polymers' composition and its ability to complex Cd, Cu and Pb". *Chemosphere*, No. 10: 1701–10. https://doi.org/10.1016/S0045-6535(03)00355-2.

Guibaud, Gilles, Sophie Comte, François Bordas, Séverine Dupuy, and Michel Baudu. 2005. "Comparison of the complexation potential of extracellular polymeric substances (EPS), extracted from activated sludges and produces by pure bacteria strains, for cadmium, lead and nickel". *Chemosphere*, No. 5: 629–38. https://doi.org/10.1016/j.chemosphere.2004.10.028.

Guiné, Véronique, Jean M. F. Martins, B. Causse, A. Durand, Jean-Paul Gaudet, and Lorenzo Spadini. 2007. "Effect of cultivation and experimental conditions on the surface reactivity of the metal-resistant bacteria *Cupriavidus metallidurans* CH34 to protons, cadmium and zinc". *Chemical Geology*, No. 3: 266–80. https://doi.org/10.1016/j.chemgeo.2006.10.001.

Guiné, Véronique, Lorenzo Spadini, Géraldine Sarret, Myriam Muris, Cécile Delolme, Jean-Paul Gaudet, and Jean M. F. Martins. 2006. "Zinc sorption to three gram-negative bacteria: Combined titration, modeling, and EXAFS study". *Environmental Science & Technology*, No. 6: 1806–13. https://doi.org/10.1021/es0509811.

Haferburg, Gotz, and Erika Kothe. 2007. "Microbes and metals: Interactions in the environment". *Journal of Basic Microbiology*, No. 6: 453–67. https://doi.org/10.1002/jobm.200700275.

Ismail, Ibrahim, and Tarek Moustafa. 2016. "Chapter: Biosorption of heavy metals". In *Heavy Metals: Sources, Toxicity and Remediation Techniques*, edited by Deepak Pathania, 131–74. New York: Nova Science Publishers.

Keasling, Jay D., Stephen J. Van Dien, and Jaya Pramanik. 1998. "Engineering polyphosphate metabolism in *Escherichia coli*: Implications for bioremediation of inorganic contaminants". *Biotechnology and Bioengineering*, No. 2–3: 231–9. https://doi.org/10.1002/(sici)1097-0290(19980420)58:2/3<231::aid-bit16>3.0.co;2-f.

Kelly, S. D., Maxim I. Boyanov, Bruce A. Bunker, Jeremy B. Fein, David A. Fowle, N. Yee, and K. M. Kemner. 2001. "XAFS determination of the bacterial cell wall functional groups responsible for complexation of Cd and U as a function of pH". *Journal of Synchrotron Radiation*, No. 8: 946–8. https://doi.org/10.1107/S0909049500021014.

Kenney, Jeremy P. L., and Janice B. Fein. 2011. "Importance of extracellular polysaccharides on proton and Cd binding to bacterial biomass: A comparative study". *Chemical Geology*, No. 3–4: 109–17. https://doi.org/10.1016/j.chemgeo.2011.04.011.

Keunen, Els, Sascha Truyens, Liesbeth Bruckers, Tony Remans, Jaco Vangronsveld, and Ann Cuypers. 2011. "Survival of Cd-exposed *Arabidopsis thaliana*: Are these plants reproductively challenged?". *Plant Physiology and Biochemistry*, No. 10: 1084–91. 10.1016/j.plaphy.2011.07.013.

Kratochvil, David, and Bohumil Volesky. 1998. "Advances in the biosorption of heavy metals". *Trends in Biotechnology*, No. 7: 291–300. https://doi.org/10.1016/S0167-7799(98)01218-9.

Ledin, Maria. 2000. "Accumulation of metals by microorganisms – Processes and importance for soil systems". *Earth-Science Reviews*, No. 1–4: 1–31. https://doi.org/10.1016/S0012-8252(00)00008-8.

Lee, Garam, Zhaxi Suonana, Seung Hyeon Kim, Dong-Woon Hwang, and Kun-Seop Lee. 2019. "Heavy metal accumulation and phytoremediation potential by transplants of the seagrass *Zostera marina* in the polluted bay systems". *Marine Pollution Bulletin*, No. 149: 110509. https://doi.org/10.1016/j.marpolbul.2019.110509

Lee, Sun Mi, Gregor Grass, Christopher J. Haney, Bin Fan, Barry P. Rosen, Andreas Anton, Dietrich H. Nies, and Christopher Rensing. 2002. "Functional analysis of the *Escherichia coli* zinc transporter ZitB". *FEMS Microbiology Letters*, No. 2: 273–8. https://doi.org/10.1111/j.1574-6968.2002.tb11402.x.

Limmer, Matt, and Joel Burken. 2016. "Phytovolatilization of organic contaminants". *Environmental Science and Technology*, No. 50: 6632–43. https://doi.org/10.1021/acs.est.5b04113.

Ling, Tao, Qiang Gao, Haolin Du, Qiancheng Zhao, and Jun Ren. 2017. "Growing, physiological responses and Cd uptake of Corn (*Zea mays* L.) under different Cd supply". *Chemical Speciation & Bioavailability*, No. 1: 216–21. https://doi.org/10.1080/09542299.2017.1400924.

Luo, Zhi-Bin, Jiali He, Andrea Polle, and Heinz Rennenberg. 2016. "Heavy metal accumulation and signal transduction in herbaceous and woody plants: Paving the way for enhancing phytoremediation efficiency". *Biotechnology Advances*, No. 6: 1131–48. https://doi.org/10.1016/j.biotechadv.2016.07.003.

Madanan, Minisha Thalikulangara, Irfan Khursheed Shah, George K. Varghese, and Rajendra K. Kaushal. 2021. "Application of Aztec Marigold (*Tagetes erecta* L.) for phytoremediation of heavy metal polluted lateritic soil". *Environmental Chemistry and Ecotoxicology*, N0. 3: 17–22. https://doi.org/10.1016/j.enceco.2020.10.007.

Mahajan, Pooja, and Jyotsna Kaushal. 2018. "Role of phytoremediation in reducing cadmium toxicity in soil and water". *Journal of Toxicology*, No. 3: 1–16. https://doi.org/10.1155/2018/4864365.

Malik, Anushree. 2004. "Metal bioremediation through growing cells". *Environment International*, No. 2: 261–78. 10.1016/j.envint.2003.08.001.

Manoj, Srinivas Ravi, Chinnannan Karthik, Krishna Kadirvelu, Padikasan Indra Arulselvi, Thangavel Shanmugasundaram, Benedict Bruno, and Mani Rajkumar. 2020. "Understanding the molecular mechanisms for the enhanced phytoremediation of heavy metals through plant growth promoting rhizobacteria: A review". *Journal of Environmental Management*, No. 254: 109779. https://doi.org/10.1016/j.jenvman.2019.109779.

Martin, William F. 2019. "Carbon–metal bonds: Rare and primordial in metabolism". *Trends in Biochemical Sciences*, No. 9: 807–18. https://doi.org/10.1016/j.tibs.2019.04.010.

Meda, Anderson R., Enrico B. Scheuermann, Ulrich E. Prechsl, Bülent Erenoglu, Gabriel Schaaf, Heiko Hayen, Günther Weber, and Nicolaus von Wirén. 2007. "Iron acquisition by phytosiderophores contributes to cadmium tolerance". *Plant Physiology*, No. 4: 1761–73. https://doi.org/10.1104/pp.106.094474.

Mergeay, Max, D. H. Nies, H. G. Schlegel, Joep Gerits, P. Charles, and Frédérique Van Gijsegem. 1985. "*Alcaligenes eutrophus* CH34 is a facultative chemolitotroph with plasmid-bound resistance to heavy metal". *Journal of Bacteriology*, No. 1: 328–34.

Milner, Mattew J., and Leon V. Kochian. 2008. "Investigating heavy-metal hyperaccumulation using *Thlaspi caerulescens* as a model system". *Annals of Botany*, No. 1: 3–13. https://doi.org/10.1093/aob/mcn063.

Mishra, Bhoopesh, Maxim I. Boyanov, Bruce A. Bunker, S. D. Kelly, K. M. Kemner, Robert Nerenberg, Brenda L. Read-Daily, and Jeremy B. Fein. 2009. "An X-ray absorption spectroscopy study of Cd binding onto bacterial consortia". *Geochimica et Cosmochimica Acta*, No. 15: 4311–25. https://doi.org/10.1016/j.gca.2008.11.032.

Mugwar, Ahmed J., and Michael J. Harbottle. 2016. "Toxicity effects on metal sequestration by microbially-induced carbonate precipitation". *Journal of Hazardous Materials*, No. 314: 237–48. https://doi.org/10.1016/j.jhazmat.2016.04.039.

Neumann, D., O. Lichtenberger, D. Gunther, K. Tschiersch, and L. Nover. 1994. "Heat-shock proteins induce heavy-metal tolerance in higher plants". *Planta*, No. 3: 360–7. https://doi.org/10.1007/BF00197536.

Ngwenya, Bryne T., Ian W. Sutherland, and Lynn Kennedy. 2003. "Comparison of the acid-base behaviour and metal adsorption characteristics of a gram-negative bacterium with other strains". *Applied Geochemistry*, No. 4: 527–38. https://doi.org/10.1016/S0883-2927(02)00118-X.

Nies, Dietrich H. 1999. "Microbial heavy-metal resistance". *Applied Microbiology and Biotechnology*, No. 51: 730–50. https://doi.org/10.1007/s002530051457.

Nies, Dietrich H. 2003. "Efflux-mediated heavy metal resistance in prokaryotes". *FEMS Microbiology Reviews*, No. 27: 313–39. 10.1016/S0168-6445(03)00048-2.

Nies, Dietrich H., and Simon Silver. 1995. "Ion efflux systems involved in bacterial metal resistances". *Journal of Industrial Microbiology*, No. 2: 186–99. https://doi.org/10.1007/BF01569902.

Nouha, Klai, Ram Saurabh Kumar, Sellamuthu Balasubramanian, and Rajeshwar DayalTyagi. 2018. "Critical review of EPS production, synthesis and composition for sludge flocculation". *Journal of Environmental Sciences*, No. 66: 225–45. https://doi.org/10.1016/j.jes.2017.05.020.

Page, Valérie, Laure Weisskopf, and Urs Feller. 2006. "Heavy metals in white lupin: Uptake, root-to-shoot transfer and redistribution within the plant". *New Phytologist*, No. 2: 329–34. https://doi.org/10.1111/j.1469-8137.2006.01756.x.

Pagès, Delphine, Lisa Sanchez, Sandrine Conrod, Xavier Gidrol, Agnes Fekete, Ph. Schmitt-Kopplin, Thierry Heulin, and Wafa Achouak. 2007. "Exploration of intraclonal adaptation mechanisms of *Pseudomonas brassicacearum* facing cadmium toxicity". *Environmental Microbiology*, No. 11: 2820–35. https://doi.org/10.1111/j.1462-2920.2007.01394.x.

Pal, Arundhati, and A. K. Paul. 2008. "Microbial extracellular polymeric substances: Central elements in heavy metal bioremediation". *Indian Journal of Microbiology*, No. 1: 49–64. https://doi.org/10.1007/s12088-008-0006-5.

Peñuela, Jorge, José Daniel Martínez, Mary Lorena Araujo, Felipe Brito, Giuseppe Lubes, Mildred Rodríguez, and Vito Lubes. 2011. "Speciation of the nickel(II) complexes with oxalic and malonic acids studied in $1.0 \, mol \, dm^{-3}$ NaCl at 25°C". *Journal of Coordination Chemistry*, No. 15: 2698–705. https://doi.org/10.1080/0 0958972.2011.605443.

Plateau, Pierre, Cosmin Saveanu, Roxane Lestini, Marc Dauplais, Laurence Decourty, Alain Jacquier, Sylvain Blanquet, and Myriam Lazarda. 2017. "Exposure to selenomethionine causes selenocysteine misincorporation and protein aggregation in *Saccharomyces cerevisiae*". *Scientific Reports*, No. 7: 44761. https://doi.org/10.1038/srep44761.

Reese, Ralph Neil, and Dennis R. Winge. 1988. "Sulfide stabilization of cadmium γ-glutamyl peptide complex of *Schizosaccharomyces pombe*". *Journal of Biological Chemistry*, No. 26: 12832–5.

Reese, Ralph Neil, and G. J. Wagner. 1987. "Properties of tobacco cadmium-binding peptide(s)". *Biochemical Journal*, No. 3: 641–7. https://doi.org/10.1042/bj2410641.

Reese, Ralph Neil, Cindy A. White, and Dennis R. Winge. 1992. "Cadmium-sulfide crystallites in Cd-(γ-EC)$_n$G peptide complexes from tomato". *Plant physiology*, No. 1: 225–9. https://doi.org/10.1104/pp.98.1.225.

Reichman, Suzie M., and David R. Parker. 2005. "Chapter 4 – Metal complexation by phytosiderophores in the rhizosphere". In *Biogeochemistry of Trace Elements in the Rhizosphere*, edited by P. M. Huang and G. R. Gobran, 129–56. Cambridge, MA: Academic Press Elsevier Inc. https://doi.org/10.1016/B978-044451997-9/50006-4.

Richau, Kerstin H., and Henk Schat. 2009. "Intraspecific variation of nickel and zinc accumulation and tolerance in the hyperaccumulator *Thlaspi caerulescens*". *Plant Soil*, No. 1: 253–62. https://doi.org/10.1007/s11104-008-9724-z.

Rodríguez-Ortíz, Juan Carlos, Ricardo David Valdez-Cepeda, J. L. Lara-Mireles, H. Rodríguez-Fuentes, R. E. Vázquez-Alvarado, R. Magallanes-Quintanar, and G. Hernández. 2006. "Soil nitrogen fertilization effects on phytoextraction of cadmium and lead by tobacco (*Nicotiana tabacum* L.)". *Bioremediation Journal*, No. 3: 105–14. https://doi.org/10.1080/10889860600939815.

Roux, Murielle, Géraldine Sarret, Isabelle Pignot-Paintrand, Marc Fontecave, and Jacques. 2001. "Mobilization of selenite by *Ralstonia metallidurans* CH34". *Applied and Environmental Microbiology*, No. 2: 769–73. https://doi.org/10.1128/AEM.67.2.769-773.2001.

Saier, M. H. Jr., R. Tam, Aiala Reizer, and J. Reizer. 1994. "Two novel families of bacterial membrane proteins concerned with nodulation, cell division and transport". *Molecular Microbiology*, No. 5: 841–7. https://doi.org/10.1111/j.1365-2958.1994.tb00362.x.

Salt, David E., Roger C. Prince, Ingrid J. Pickering, and Ilya Raskin. 1995. "Mechanisms of cadmium mobility and accumulation in Indian mustard". *Plant Physiology*, No. 4: 1427–33. https://doi.org/10.1104/pp.109.4.1427.

Sand, Wolfganag, and Tilman Gehrke. 2006. "Extracellular polymeric substances mediate bioleaching/biocorrosion via interfacial processes involving iron(III) ions and acidophilic bacteria". *Research in Microbiology*, No. 157: 49–56. https://doi.org/10.1016/j.resmic.2005.07.012.

Sarret, Géraldine, Laure Avoscan, Marie Carriere, Richard N. Collins, Nicolas Geoffroy, Francine Carrot, Jacques Covès, and Barbara Gouget. 2005. "Chemical forms of selenium in the metal-resistant bacterium *Ralstonia metallidurans* CH34 exposed to selenite and selenate". *Applied and Environmental Microbiology*, No. 5: 2331–7. 10.1128/AEM.71.5.2331-2337.2005.

Sarret, Géraldine, Pierre Saumitou-Laprade, Valérie Bert, Olivier Proux, Jean-Louis Hazemann, Agnès Traverse, Matthew A. Marcus, and Alain Manceau. 2002. "Forms of zinc accumulated in the hyperaccumulator *Arabidopsis halleri*". *Plant Physiology*, No. 4: 1815–826. https://doi.org/10.1104/pp.007799.

Schaaf, Gabriel, Emin Bülent Erenoglu, and Nicolaus von Wirén. 2004. "Physiological and biochemical characterization of metal-phytosiderophore transport in graminaceous species". *Soil Science and Plant Nutrition*, No. 7: 989–95. https://doi.org/10.1080/00380768.2004.10408565.

Shackira, A. M., and J. T. Puthur. 2019. "Phytostabilization of heavy metals: Understanding of principles and practices". In *Plant-Metal Interactions*, edited by Sudhakar Srivastava, Ashish K. Srivastava, and Penna Suprasanna, 263–82. Cham, Switzerland: Springer. https://doi.org/10.1007/978-3-030-20732-8_13.

Shahabaldin, Rezania, Mat Taib Shazwin, Fadhil Md Dina Mohd, Aini Dahalan Farrah, and Kamyab Hesam. 2016. "Comprehensive review on phytotechnology: Heavy metals removal by diverse aquatic plants species from wastewater: Review". *Journal of Hazardous Materials*, No. 318: 587–99. https://doi.org/10.1016/j.jhazmat.2016.07.053.

Sidhu, Gagan Preet Singh, Aditi Shreeya Bali, and Renu Bhardwaj. 2019. "Role of organic acids in mitigating cadmium toxicity in plants". In *Cadmium Tolerance in Plants: Agronomic, Molecular, Signaling and Omics Approaches*, edited by Kamrun Nahar, Mirza Hasanuzzaman, and Majeti Narasimha Vara Prasad. Cambridge, MA: Academic Press Elsevier. https://doi.org/10.1016/B978-0-12-815794-7.00010-2.

Song, Yu, Liang Jin, and Xiaojuan Wang. 2016. "Cadmium absorption and transportation pathways in plants". *International Journal of Phytoremediation*, No. 2: 133–41. https://doi.org/10.1080/15226514.2016.1207598.

Sors, Thomas G., Catherine P. Martin, and David E. Salt. 2009. "Characterization of selenocysteine methyltransferases from *Astragalus* species with contrasting selenium accumulation capacity". *The Plant Journal*, No. 1: 110–22. https://doi. org/10.1111/j.1365-313X.2009.03855.x.

Steliga, Teresa, and Dorota Kluk. 2020. "Application of *Festuca arundinacea* in phytoremediation of soils contaminated with Pb, Ni, Cd and petroleum hydrocarbons". *Ecotoxicology and Environmental Safety*, No. 194: 110409. https://doi. org/10.1016/j.ecoenv.2020.110409.

Suman, Jáchym, Ondrej Uhlik, Jitka Viktorova, and Tomas Macek. 2018. "Phytoextraction of heavy metals: A promising tool for clean-up of polluted environment". *Frontiers in Plant Science*, No. 9: 1476. https://doi.org/10.3389/fpls.2018.01476.

Ueshima, Masato, Brian R. Ginn, Elizabeth A. Haack, Jennifer E. S. Szymanowski, and Jeremy B. Fein. 2008. "Cd adsorption onto *Pseudomonas putida* in the presence and absence of extracellular polymeric substances". *Geochimica et Cosmochimica Acta*, No. 24: 5885–95. https://doi.org/10.1016/j.gca.2008.09.014.

Uroz, Stephane, Christophe Calvaruso, Marie-Pierre Turpault, and Pascale Frey-Klett. 2009. "Mineral weathering by bacteria: Ecology, actors and mechanisms". *Trends in Microbiology*, No. 8: 378–87. https://doi.org/10.1016/j.tim.2009.05.004.

Van Steveninck, R. F. M., Margaret E. Van Steveninck, Denise R. Fernando, Walter J. Horst, and H. Marschner. 1987. "Deposition of zinc phytate in globular bodies in roots of *Deschampsia caespitosa* ecotypes; a detoxification mechanism?". *Journal of Plant Physiology*, No. 3–4: 247–57. https://doi.org/10.1016/S0176-1617(87)80164-5.

Volesky, B., and Z. R. Holan. 1995. "Biosorption of heavy metals". *Biotechnology Progress*, No. 3: 235–50. https://doi.org/10.1021/bp00033a001.

Wingender, Jost, Thomas R. Neu, and Hans-Curt Flemming. 1999. "What are bacterial extracellular polymeric substances?". In *Microbial Extracellular Polymeric Substances: Characterization, Structure, and Function*, edited by Jost Wingender, Thomas R. Neu, and Hans-Curt Flemming, 1–19. Berlin and New York: Springer. https://doi.org/10.1007/978-3-642-60147-7_1.

Worms, Isabelle A. M., Dana F. Simon, Christel S. Hassler, and Kevin J. Wilkinson. 2006. "Bioavailability of trace metals to aquatic microorganisms: Importance of chemical, biological and physical processes on biouptake". *Biochimie*, No. 11: 1721–31. https://doi.org/10.1016/j.biochi.2006.09.008.

Xie, Yan, Longxing Hu, Zhimin Du, Xiaoyan Sun, Erick Amombo, Jibiao Fan, and Jinmin Fu. 2014. "Effects of cadmium exposure on growth and metabolic profile of bermudagrass [*Cynodon dactylon* (L.) Pers.]". *PLoS ONE*, No. 12: e115279. https://doi.org/10.1371/journal.pone.0115279.

Xu, Zhaolong, Xiaoqing Liu, Xiaolan He, Ling Xu, Yihong Huang, Hongbo Shao, Dayong Zhang, Boping Tang, and Hongxiang Ma. 2017. "The soybean basic helix-loop-helix transcription factor org3-like enhances cadmium tolerance via increased iron and reduced cadmium uptake and transport from roots to shoots". *Frontiers in Plant Science*, No. 8: 1098. https://doi.org/10.3389/fpls.2017.01098.

Yang, Yang, Yichen Ge, Hongyuan Zeng, Xihong Zhou, Liang Peng, and Qingru Zeng. 2017. "Phytoextraction of cadmium-contaminated soil and potential of regenerated tobacco biomass for recovery of cadmium". *Scientific Reports*, No. 1: 7210. https://doi.org/10.1038/s41598-017-05834-8.

Yao, Xiani, Yuerong Cai, Diqiu Yu, and Gang Liang. 2018. "bHLH104 confers tolerance to cadmium stress in *Arabidopsis thaliana*". *Journal of Integrative Plant Biology*, No. 8: 691–702. https://doi.org/10.1111/jipb.12658.

Yee, Nathan, and Jeremy B. Fein. 2001. "Cd adsorption onto bacterial surfaces: A universal adsorption edge?". *Geochimica et Cosmochimica Acta*, No. 13: 2037–42. https://doi.org/10.1016/S0016-7037(01)00587-7.

Yoon, Kyong Pyo, and S. Silver. 1991. "A second gene in the *Staphylococcus aureus cadA* cadmium resistance determinant of plasmid pI258". *Journal of Bacteriology*, No. 23: 7636–42. 10.1128/jb.173.23.7636-7642.1991.

Zanella, Letizia, Laura Fattorini, Patrizia Brunetti, Enrica Roccotiello, Laura Cornara, Simone D'Angeli, Federica Della Rovere, Maura Cardarelli, Maurizio Barbieri, Luigi Sanità di Toppi, Francesca Degola, Sylvia Lindberg, Maria Maddalena Altamura, and Giuseppina Falasca. 2016. "Overexpression of AtPCS1 in tobacco increases arsenic and arsenic plus cadmium accumulation and detoxification". *Planta*, No. 243: 605–22. https://doi.org/10.1007/s00425-015-2428-8.

Zwolak, Iwona. 2020. "The role of selenium in arsenic and cadmium toxicity: An updated review of scientific literature". *Biological Trace Element Research*, No. 193: 44–63. https://doi.org/10.1007/s12011-019-01691-w.

5 Approaches for increasing bioremediation capabilities of plants and microorganisms towards heavy metals and radionuclides

Introduction

Genetic manipulation can constitute another way of developing efficient plants and microorganisms for bioremediation of heavy metals and radionuclides.

In order to develop organisms capable of supporting high concentrations of heavy metals and radionuclides, it is essential to understand the molecular and biochemical mechanisms adopted by organisms to resist the toxic effects of heavy metals and radionuclides. Furthermore, synergistic use of efficient microbes and suitable plants can significantly boost remediation efficiency (Agnihotri and Seth 2019, 240; Kazemalilou, Delangiz, Lajayer, and Ghorbanpour 2020, 130).

Improving phytoremediation with biotechnology

Very few plants have a high endurance in metal-contaminated soils and a high biomass productivity, like Moso bamboo, *Phyllostachys praecox* (Bian, Zhong, Zhang, Yang, et al. 2020). So, to improve plant performance, some methods were proposed.

Use of complexing or chelating reagents

The efficiency of depollution in the soil–plant system depends on the plant's productivity in terms of accumulation of metals and biomass production and on the solubility of metals in the ground.

The conventional use of hyperaccumulator plants with low biomass generally does not achieve high extraction yields. In addition, hyperaccumulative plants often target only one metal, so they are not suitable for treating polymetallic pollution.

DOI: 10.4324/9781003282600-6

Current phytodecontamination strategies involve the use of metallotolerant plants with large biomass and an average accumulation capacity. The objective is to increase metals' mobility in the soil to increase their accumulation in the plant.

The interactions existing in the soil–plant system, particularly at the rhizosphere level, have the consequence of modifying the speciation and the mobility of metals by making them available to the plant. At the root level, the excretion of organic molecules of low molecular weight, such as acetic, oxalic, fumaric, and citric acids, allows the formation of soluble complexes between these molecules and the metal. The use of complexing or chelating reagents to improve the efficiency of phytoextraction is based on this observation (Chen, Lin, Luo, He, et al. 2003, 807; Keeling, Stewart, Anderson, and Robinson 2003, 243; Hauser, Tandy, Schulin, and Nowack 2005, 6819). Natural chelators (such as citric and oxalic acids) or synthetic chelators (such as Ethylene Diamine Tetra-Acetic (EDTA), Diethylenetriamine Pentaacetic Acid (DTPA), and NitriloTriAcetic acid (NTA)), having a strong affinity for metallic elements, are used to dissolve metals in the soil. Indeed, the complexing agents compete with the constituents of the soil with which the metals are associated. Part of the metal stock then goes into solution in the form of stable and soluble metal–chelate complexes (Norvell 1984, 1285; Peters 1999, 151).

Numerous phytoextraction studies of different metals, such as Pb, Zn, Cu, Cd, Ni, Co, and gold (Au), in the presence of chelating agents have been carried out. The results obtained are very variable and depend on the target metal, the chelator, and the plant used. Some of these studies show promising results: authors obtain an extraction of the metal from the soil and an accumulation of this metal in the plant's leaves. In other cases, metal build-up by the plant is present, but results were in loss of biomass or damage to the plant. Thus, the accumulation of high metal contents induced by chelating agents causes plant stress and affects its productivity, especially in the case of polymetallic pollution, when several metals with different degrees of phytotoxicity are bioavailable (Huang, Chen, Berti, and Cunningham 1997, 800; Vassil, Kapulnik, Raskin, and Salt 1998, 447; Cooper, Sims, Cunningham, Huang, et al. 1999, 1709; Robinson, Brooks, and Clothier 1999, 689; Wu, Hsu, and Cunningham 1999, 1898; Kayser, Wenger, Keller, Attinger, et al. 2000, 1778; Robinson, Mills, Petit, Fung, et al. 2000, 689; Lombi, Zhao, Dunham, and McGrath 2001, 1919; Thayalakumaran, Robinson, Vogeler, Scotter, et al. 2003, 415; Thayalakumaran, Vogeler, Scotter, Percival, et al. 2003a, 323; Thayalakumaran, Vogeler, Scotter, Percival, et al. 2003b, 335; Luo, Shen, Lou, Li 2006, 862; Evangelou, Ebel, and Schaeffer 2007, 989). For example, Zn and especially Cu are more toxic than Pb for most plants. Thus, when the plant simultaneously accumulates these three metals, its growth is affected when the lowest phytotoxicity (Cu) threshold is reached (Lombi, Zhao, Dunham, and McGrath 2001, 1924). In general, chelators effectively remobilize metals in the soil and induce their accumulation in plants. The most promising results have been obtained for phytoextraction of Pb by Indian mustard (*Brassica juncea*) in the presence of EDTA (Blaylock, Salt, Dushenkov, Zakharova, et al. 1997, 860). However, it should not be forgotten that the application of chelating agents in the natural environment poses risks. Due to their strong complexing power and persistence in the environment, synthetic chelators (EDTA type) can cause massive migration of metals into the environment due to their low biodegradability. Komárek, Tlustoš, Száková, and Chrastny (2008) assessed the risks associated with phytoextraction of Pb by a hybrid poplar tree (*Populus nigra* L. × *Populus maximoviczii* Henry.) in the presence of EDTA. They have shown that the application of EDTA keeps metals, Pb and Cu, in the soil in highly mobile forms for two years after application. These metals can be leached to the aquifers or be absorbed by plants, then animals, and thus reach the food chain (Komárek, Tlustoš, Száková, and Chrastny 2008, 27). The migration of metals through the soil also has toxic effects on the microorganisms present in the soil (Grcman, Velikonja-Bolta, Vodnik, Kos, et al. 2001, 113).

Consequently, the use of chelating agents in phytoremediation requires an adapted and careful approach. First, their degradation and their toxicity must be examined beforehand. Second, their use in the natural environment requires a device to contain the leaching of the soil to avoid pollution of the aquifers following the remobilization of metals and the spillage of complexing agents. Finally, chelators can affect plant growth.

For these reasons, the use of plants with an extensive root system, comprising underground stems of the rhizome type, seems to be an interesting path, although little explored.

Plant improvement

The induction of in vitro neoformation of plants in rapeseed *Brassica napus* L. by the in vitro culture of transverse thin cell layers constitutes a fast and effective way to develop strategies for varietal improvement by biotechnological processes to obtain newly formed plants that accumulate toxic metals that can be used for phytoremediation. This technique made it possible to quickly observe the new formation of buds in rapeseed, including in the presence of large doses of Zn (up to 2000 µM) and Pb (up to 1500 µM) (Ben Ghnaya, Charles, Ben Hamida, and Branchard 2006, 82).

Selective breeding

Selective breeding between native and non-native plants species is done to develop improved clones for biomass production, phytoremediation, nutrient filters, and streambank stabilization (Wei, da Silva, and Zhou 2008, 666; Gómez, Contreras, Bolonio, Quintana, et al. 2019, 13). It is the case of the willow *Salix* in North America (Kopp, Smart, Maynard, Isebrands, et al. 2001, 287; Pulford and Watson 2003, 537) and *Arabidopsis halleri* in the North of France (Bert, Meerts, Saumitou-Laprade, Salis, et al. 2003, 9).

Transgenesis

Other phytoextraction strategies, called "genoremediation", aim to increase the amount of metal extracted during each life cycle through genetic transgenesis. Even though the extraction of the metal by the plant requires the absorption and the transfer of the metal in the harvestable parts, the ideal plant for phytoextraction must combine the capacity to accumulate high contents of metals, like *Thlaspi caerulescens*, and that to produce high biomass, such as *Brassica juncea* rapidly. Thus, several genetic approaches were developed. These studies relate to the introduction of genes (from microbes, plants, or animals) responsible for the production of proteins and peptides, allowing the accumulation and tolerance of a given metal in plants with rapid growth and large biomass (Mejare and Bulow 2001, 70; Berken, Mulholland, LeDuc, and Terry 2002, 567; Arthur, Rice, Rice, Anderson, et al. 2005, 116; Singh, Kang, Mulchandani, and Chen 2008, 442; Sarwar, Imran, Shaheen, Ishaque, et al. 2017, 716; Kaur, Yadav, Kohli, Kumar, et al. 2018, 63; Ludvíková and Griga 2019, 341; Rai, Kim, Lee, and Lee 2020).

The trend is accentuated with the development of genetically modified plants. The goal is to generate plants that can survive in highly contaminated environments and accumulate or transform larger quantities of toxic substances (Pilon-Smits and Pilon 2002, 442). For example, the potential for a hyperaccumulation of As has been successfully tested by modifying the plant mouse-ear cress *Arabidopsis thaliana* by providing genes coding for arsenate reductase and γ-glutamylcysteine synthetase (Dhankher, Li, Rosen, Shi, et al. 2002, 1140). The potential of resistance to both Ni and Pb has been successfully made by modifying a tobacco plant (*Nicotiana tabacum* var. Samsun) by providing genes coding for calmodulin-binding transporter (Arazi, Sunkar, Kaplan, and Fromm 1999, 171). Mouse-ear cress plants (*A. thaliana*) and tobacco plants (*N. tabacum*) engineered with a modified bacterial mercuric reductase gene (*merA*) show promise as tools for mercury phytoremediation (Heaton, Rugh, Wang, and Meagher 1998, 497). Tobacco plants (*N. tabacum*) expressing the *Arabidopsis* antiporter *CAX2* (calcium exchanger 2) accumulated more Ca^{2+}, Cd^{2+}, and Mn^{2+} (Hirschi, Korenkov, Wilganowski, and Wagner 2000, 125).

Improving microbial remediation with biotechnology

The extraordinarily rapid advances in molecular biology have increased the efficiency of microbial bioremediation. For example, the insertion of the *ArsR* gene into the bacterium *Escherichia coli* allows it to accumulate 60 times of As than its unmodified counterpart (Kostal, Yang, Wu, Mulchandani, et al. 2004, 4582). Efficiency can also be boosted by inserting, into this same bacterium, the genes (isolated from the yeast *Schizosaccharomyces pombe*) of phytochelatin synthase and of γ-glutamylcysteine synthetase, and the *GlpF* gene, which codes for an As uptake transporter (Singh, Kang, Lee, Mulchandani, et al. 2010, 780).

The use of genetically modified strains can be facilitated when they are developed to make it impossible to transfer recombinant genes to other microorganisms and/or to drastically confine the genetically modified strains in the site to be treated and prevent any release into the environment. The basic strategies for achieving these objectives were described in 1993 by Molin, Boe, Jensen, Kristensen, and colleagues. The principle is, in theory, straightforward: it consists in triggering the expression of a gene qualified as "suicide" because its product kills the cell which contains it, when this one is no longer in the presence of the compounds to be degraded or when a plasmid carrying foreign genes has been transferred by another bacterium (Molin, Boe, Jensen, Kristensen, et al. 1993, 139).

Contreras, Molin, and Ramos (1991) were the first to apply this principle to build a containable strain capable of digesting polluting substances. For this, they had used the regulation circuit of the expression of the degradation genes carried by the TOL plasmid pWW0 from *Pseudomonas putida*. The presence of many types of benzoates (alkyl- and halo-substituted) in the environment of bacteria causes the expression of the regulatory gene *xylS*, the product of which activates the genes coding for the enzymes of the degradation pathway of these benzoates. Thus, the authors merged the *lacI* gene of *Escherichia coli* (which codes for the repressor of the lactose operon (LACI) to a promoter activated by the product of the *xylS* gene. Next to this construction, they had introduced the promoter of the lactose operon fused to the *gef* gene of *E. coli*. The latter codes for a peptide that kills cells by causing irreversible membrane damage. When the bacteria which contain this construction are in the presence of compounds that activate, via the product of the *xylS* gene (the degradation genes), the LACI is synthesized and inhibits the expression of the *gef* gene. On the other hand, when the quantity of these compounds decreases, or when the bacteria are moved towards an environment that does not contain them, the expression of the LACI decreases. This activates the suicide gene *gef* since its promoter is no longer repressed. This containment system has been used in *E. coli* and *Pseudomonas putida* in other studies but, because a relatively unstable plasmid carried the xylS gene, a significant number of recombinant bacteria escaped suicide when the pollutants were eliminated from the environment (Contreras, Molin, and Ramos 1991, 1504; Jensen, Ramos, Kaneva, and Molin 1993, 3713; Ronchel, Ramos, Jensen, Molin, et al. 1995, 2990).

Molina, Ramos, Ronchel, Molin, et al. (1998) strongly stabilized the system by cloning all of the components in the chromosome of *P. putida*. In laboratory conditions, one bacterium in 10^8 managed to escape suicide in the absence of pollutants. The authors also used the strain in external tests (with the authorization of the Spanish Ministry of the Environment). They found that the containable bacteria could colonize the rhizosphere of plants only in soils contaminated with 3-methylbenzoate. They did not observe any dissemination of the strain outside the experimental area. However, the elimination rate of the recombinant bacteria was much slower under natural conditions than in the laboratory. In addition, the survival rate of 10^{-8} was still too high (Molina, Ramos, Ronchel, Molin, et al. 1998, 2072–7). Ronchel and Ramos (2001) sought to improve the suicide system by using not one but two functions capable of killing cells in the absence of pollutants. The authors introduced the entire *gef* system described earlier into a *P. putida* mutant whose chromosomal *asd* gene had been deleted. The absence of this gene has resulted in the inability to synthesize certain amino acids and diaminopimelic acid (a substance absent from the natural environment but necessary for bacteria). A construct composed of the *asd* gene (deleted from its own promoter) and a promoter activated in the presence of pollutants of benzoate type, once again via the product of *xylS* gene, was introduced into the chromosome of the strain. In the absence of pollutants, the suicide of bacteria occurs both by the synthesis of the *gef* gene product and by the absence of diaminopimelic acid. This time, the number of survivors in the absence of the pollutant fell below the detection limit (less than one survivor for 10^9 bacteria), and the disappearance of bacteria from unpolluted soil occurred in 20 to 25 days (against 100 with the previous system) (Ronchel and Ramos 2001, 2649).

A system based on the same principle has been developed by Torres, Jaenecke, Timmis, Garcia, et al. (2003) to limit the horizontal transfer of conjugative plasmids carrying cloned information. As in the previous case, and for the same reasons, it has two suicide functions. However, these were no longer induced in the absence of pollutants but when the plasmid which carries the suicide genes enters by conjugation in a bacterium not belonging to the constructed strain. The authors introduced into a conjugate plasmid of the HB101 strain of *E. coli* the *colE3* gene coding for colicin E3, capable of killing many bacteria by destroying their 16S ribosomal RNA. In the chromosome of this strain, they introduced the gene *immE3*, which constitutively produces an antidote to colicin E3. Therefore, the strain is protected from the effects of colicin, but when the plasmid enters a receptor bacterium lacking the *imm3* gene, colicin E3 exerts its effect and kills this bacterium. In addition, the plasmid carries a gene encoding the restriction enzyme EcoRI. This enzyme cuts all DNA when it recognizes a particular sequence of six base pairs. The strain of *E. coli* carries in its chromosome the gene coding for the corresponding modification enzyme, methylase, which protects the DNA sequences specifically recognized by EcoRI by methylating them. The DNA of the *E. coli* strain is therefore not cleaved by EcoRI. On the other hand, when the plasmid is introduced into a bacterium devoid of methylase, EcoRI destroys its DNA almost instantaneously. The frequency of transfer of the plasmid carrying this construct was reduced by a factor of 10^8 compared to the transfer of the same plasmid lacking the suicide system. This latest study describes the construction and evaluation of a laboratory strain. The use of this type of construction in natural conditions, however, raises two questions. The first concerns the expression of suicide genes in bacteria belonging to species other than *E. coli*. The promoters of the two genes are promoters of *E. coli*: their expression in other genetic contexts is not a certainty. The recipient bacteria which do not express them thus escape suicide. The second question relates to the dissemination of suicide genes by plasmids conjugative. The confinement of cloned genetic information would be drastic, but it would be done by the physical elimination of all bacteria capable of receiving these plasmids. The diversity of the environment concerned could be seriously affected (Torres, Jaenecke, Timmis, Garcia, et al. 2003, 3595–600).

Another experiment showed that a transgenic *Nicotiana langsdorffii* created by incorporating a *rat/rolC* gene from *Agrobacterium rhizogenes* enhanced its tolerance to Cr (Malandrino, Giacomino, Karthik, Zelano, et al. 2017, 87).

An alternative method is the use of genetically modified bacteria cells killed before they come into contact with the polluted matrix. Because DNA is stable in the environment, dead cells can transfer plasmids to other organisms. In addition, even if such cells obviously cannot multiply in the environment, they could release enzymes there or carry on their surface molecules capable of degrading or absorbing pollutants. However, their use would require developing effective methods of contacting the pollutant within the contaminated matrix and repeating them throughout the treatment. For the first time in the United States, a team from the University of Minnesota at St Paul used a recombinant strain of *E. coli* manipulated to overexpress an atrazine chlorohydrolase to decontaminate soil polluted by atrazine, a herbicide widely used in agriculture. Before their application to the surfaces to be treated, the bacteria had been killed by a chemical agent, glutaraldehyde. In eight weeks, the authors found a reduction in the concentration of atrazine by 77%. They consider this to be the minimum efficiency of their technique because it was applied in late fall when temperatures were already low enough. These authors also noted that the enzymatic activity was preserved for several months after treating the bacteria with glutaraldehyde. This stability is undoubtedly one of the keys to the effectiveness of the treatment (Strong, McTavish, Sadowsky, and Wackett 2000, 91).

The use of killed cells is a strategy that could also prove to be advantageous for the bioremediation of low-volume liquid effluents by passage through filters containing these microbial cells. This type of application would use cells carrying on their surface either enzymes that remain active after cell death or determinants capable of adsorbing at high-affinity toxic metals, radioactive or not, or other pollutants (Wu, Mulchandani, and Chen 2008, 181).

A group of researchers has reported using mycelium from certain strains of fungi (*Neurospora crassa*, *Trichoderma viridae*, *Mucor racemosus*, *Rhizopus chinensis*, *Penicillium citrinum*, *Aspergillus niger*, and *Aspergillus flavus*) to fix the ^{60}Co from contaminated effluents. During these treatments, the cells were killed by the radioactivity of the effluent but retained their capacity for sorption and desorption of a whole series of metals (Rashmi, Thommandru, Mohan, Balaji, et al. 2004, 1).

From the early 2000s, researchers had used the properties of conjugative plasmids to disseminate genes of interest in a polluted environment. These studies, carried out in microcosms without contact with the outside, made it possible to note the spontaneous transfer of plasmids carrying the degradation genes of 2,4-dichlorophenoxyacetate (2,4-D) towards receptor bacteria present in the medium and the effectiveness of the system in degrading this pollutant. Therefore, the aim of this work is no longer to disseminate organisms containing foreign genes but to spread these genes by a natural way of transferring genetic information. The bacteria receiving these conjugative plasmids into which the foreign genes have been cloned in turn become genetically modified organisms since these plasmids were constructed in vitro before transfer into the strain (Dejonghe, Goris, El Fantroussi, Höfte, et al. 2000, 3297; Newby, Gentry, and Pepper 2000, 3399).

Bioremediation (rhizosphere)/phytoremediation coupling

Roots of plants can be considered as biological soil engineers. Indeed, the latter create and maintain their environment, not only by their physical presence but also by their activity. For example, by processes like the exudation of organic acids and enzymes in the rhizosphere, the roots can allow microbial communities to maintain (Curl and Truelove 1986, 66) or increase minerals erosion (Hinsinger, Jaillard, and Dufey 1992, 977). Thus, despite the small volume of the rhizosphere in soils, it plays a central role in maintaining the soil–plant system (Gobran, Clegg, and Courchesne 1998, 107).

Among the microorganisms found in the soil, some live in symbiosis or mutualism with plants. The supply of water and nutrients to the plant, in exchange for carbonaceous substances and physical protection to microorganisms, can define these exchanges. In these associations, microorganisms can be classified into two categories:

- The ectosymbionts constitute the microorganisms colonizing the rhizosphere (outside of the root) or the rhizoplane (surface of the roots). Among these organisms are bacteria such as *Pseudomonas, Azotobacter, Bacillus, Enterobacter*, and fungi (*Trichoderma*). Bacteria of this type are then defined as rhizobacteria or rhizoplane bacteria;
- The endosymbionts group together the microorganisms living inside the cells of the host plant. The plasmalemma can also be pierced, this being the work of so-called endophyte bacteria. These bacteria, mainly belonging to the genera *Rhizobium* and *Frankia*, can form root nodules found on the roots of legumes and woody plants.

(Gray and Smith 2005, 395; Lugtenberg and Kamilova 2009, 541–51)

Because of their rhizospheric and endophytic characteristics, bacteria have received attention in recent years to promote the establishment of plants in unfavourable conditions by promoting phytoremediation processes. These bacteria can be isolated from plants living on soils contaminated with metals or organic compounds (Chen, Xu, Zeng, Yang, et al. 2015, 745; Kong and Glick 2017, 327).

Bioaugmentation/phytoremediation coupling

Inoculation of specific microorganisms (bio-increase) in porous matrices (soils, sediments) can be used to increase the stock of metals available for the plants used for phytoextraction (with microorganisms producing siderophores, for example). As a result, the growth of inoculated microorganisms is enhanced by the supply of nutrients exuded from the roots of plants used for phytoextraction (Kazemalilou, Delangiz, Lajayer, and Ghorbanpour 2020, 125).

Energy recovery from biomass/phytoextraction coupling

In 2003, the Federal Ministry of Research in Germany launched a programme to set up an interdisciplinary network on the theme "Renewable energy from biomass from phytoextraction of contaminated soil". This programme was piloted by the Clausthaler Umwelttechnik-Institut (CUTEC Institute), Germany, for two years. Within the framework of this programme, nine projects have been carried out which aim to achieve the following objectives:

1   the establishment of a network of experts making it possible to assess the state of science and technology in the use of biomass from soils decontaminated by phytoextraction,
2   the evaluation of the possibilities and limits of this process,
3   the design of priority projects.

This network was made up of 35 experts from various scientific and industrial sectors who work on themes such as

- recovery of heavy metals from plants;
- the different techniques currently available;
- the best methods for the energetic use of plants (Ex: combustion);
- the limits of current processes and possible new solutions.

(Nouri and Haddioui 2016, 54)

Animal remediation/phytoremediation coupling

Animal remediation depends on the characteristics of certain lower animals adsorbing and degrading heavy metals, thus eliminating and inhibiting their toxicity. Earthworms "ecosystem engineers" constitute a significant component of the soil macrofauna in most terrestrial ecosystems. Earthworms increase the availability of heavy metals in certain situations and help maintain soil structure and quality. In addition, the introduction of earthworms into soils contaminated with metals has been suggested to aid phytoremediation processes (Lemtiri, Liénard, Alabi, Brostaux, et al. 2016, 67).

Animal remediation/bioremediation coupling

By stimulating microbial activities, earthworms improve the breakdown of organic matter, leading to an increase in the content of low and high molecular weight organic acids. This process tends to decrease the availability of certain metals by incorporating metallic elements (Pb and Cu) into the complexes of organic matter and increased the availability of Cd and Zn by forming chelates with soluble organic molecules (Zhang, Mora, Dai, Chen, et al. 2016, 65).

Reference list

Agnihotri, Ashish, and Chandra Shekhar Seth. 2019. "Chapter 11 – Transgenic Brassicaceae: A promising approach for phytoremediation of heavy metals". In *Transgenic Plant Technology for Remediation of Toxic Metals and Metalloids*, edited by Majeti Narasimha Vara Prasad, 239–55. Cambridge, MA: Academic Press Elsevier Inc. https://doi.org/10.1016/B978-0-12-814389-6.00011-0.

Arazi, Tzahi, Ramanjulu Sunkar, Boaz Kaplan, and Hillel Fromm. 1999. "A tobacco plasma membrane calmodulin-binding transporter confers Ni^{2+} tolerance and Pb^{2+} hypersensitivity in transgenic plants". *The Plant Journal*, No. 2: 171–82. 10.1046/j.1365–313x.1999.00588.x.

Arthur, Ellen L., Pamela J. Rice, Patricia J. Rice, Todd A. Anderson, Sadika M. Baladi, Keri L. D. Henderson, and Joel R. Coats. 2005. "Phytoremediation: An overview". *Critical Reviews in Plant Sciences*, No. 2: 109–22. http://doi.org/10.1080/07352680590952496.

Ben Ghnaya, Asma, Gilbert Charles, Jeannette Ben Hamida, and Michel Branchard. 2006. "Phytoremediation: *In vitro* selection of Rapeseed (*Brassica napus* L.) tolerant of toxic metals". *International Journal of Tropical Geology, Geography and Ecology*, No. 692: 69–86.

Berken, Antje, Maria M. Mulholland, Danika L. LeDuc, and Norman Terry. 2002. "Genetic engineering of plants to enhance selenium phytoremediation". *Critical Reviews in Plant Sciences*, No. 21: 567–82. https://doi.org/10.1080/0735-260291044368.

Bert, Valérie, Pierre Meerts, Pierre Saumitou-Laprade, Pietro Salis, W. Gruber, and Nathalie Verbruggen. 2003. "Genetic basis of Cd tolerance and hyperaccumulation in *Arabidopsis halleri*". *Plant and Soil*, No. 1: 9–18. https://doi.org/10.1023/A:1022580325301.

Bian, Fangyuan, Zheke Zhong, Xiaoping Zhang, Chuanbao Yang, and Xu Gai. 2020. "Bamboo – an untapped plant resource for the phytoremediation of heavy metal contaminated soils". *Chemosphere*, No. 246: 125750. https://doi.org/10.1016/j.chemosphere.2019.125750.

Blaylock, Michael J., David E. Salt, Slavik Dushenkov, Olga Zakharova, Christopher Gussman, Yoram Kapulnik, Burt D. Ensley, and Ilya Raskin. 1997. "Enhanced accumulation of Pb in Indian Mustard by soil-applied chelating agents". *Environmental Science & Technology*, No. 31: 860–5. https://doi.org/10.1021/es960552a.

Chen, Ming, Piao Xu, Guangming Zeng, Chunping Yang, Danlian Huang, and Jiachao Zhang. 2015. "Bioremediation of soils contaminated with polycyclic aromatic hydrocarbons, petroleum, pesticides, chlorophenols and heavy metals by composting: Applications, microbes and future research needs". *Biotechnology Advances*, No. 6: 745–55. https://doi.org/10.1016/j.biotechadv.2015.05.003.

Chen, Y. X., Q. Lin, Y. M. Luo, Y. F. He, S. J. Zhen, Y. L. Yu, G. M. Tian, and M. H. Wong. 2003. "The role of citric acid on the phytoremediation of heavy metal contaminated soil". *Chemosphere*, No. 50: 807–11. https://doi.org/10.1016/S0045-6535(02)00223-0.

Contreras, Asunción, Soren Molin, and Juan-Luis Ramos. 1991. "Conditional-suicide containment system for bacteria which mineralize aromatics". *Applied and Environmental Microbiology*, No. 5: 1504–8. 10.1128/AEM.57.5.1504-1508.1991.

Cooper, E. M., J. T. Sims, S. D. Cunningham, J. W. Huang, and W. R. Berti. 1999. "Chelate assisted phytoextraction of lead from contaminated soils". *Journal of Environmental Quality*, No. 28: 1709–19. https://doi.org/10.2134/jeq1999.0047 2425002800060004x.

Curl, Elroy A., and Bryan Truelove. 1986. "Root exudates". In *The rhizosphere*, edited by Curl Elroy A. and Bryan Truelove, 52–95. Heidelberg: Springer-Verlag editions. https://doi.org/10.1007/978-3-642-70722-3_3.

Dejonghe, Winnie, Johan Goris, Saïd El Fantroussi, Monica Höfte, Paul De Vos, Willy Verstraete, and Eva M. Tops. 2000. "Effect of dissemination of 2,4-dichlorophenoxyacetic acid (2,4-D) degradation plasmids on 2,4-D degradation and on bacterial community structure in two different soil horizons". *Applied and Environmental Microbiology*, No. 8: 3297–304. https://doi.org/10.1128/aem.66.8.3297-3304.2000.

Dhankher, Om Parkash, Yujing Li, Barry P. Rosen, Jin Shi, David Salt, Julie F. Senecoff, Nupur A. Sashti, and Richard B. Meagher. 2002. "Engineering tolerance and hyperaccumulation of arsenic in plants combining arsenate reductase and gamma-glutamylcysteine synthetase expression". *Nature Biotechnology*, No. 20: 1140–5. https://doi.org/10.1038/nbt747.

Evangelou, Michael W. H., Mathias Ebel, and Andreas Schaeffer. 2007. "Chelate assisted phytoextraction of heavy metals from soil. Effect, mechanism, toxicity, and fate of chelating agents". *Chemosphere*, vol. 68: 989–1003. https://doi.org/10.1016/j.chemosphere.2007.01.062.

Gobran, Goerges R., S. Clegg, and François Courchesne. 1998. "Rhizospheric processes influencing the biogeochemistry of forest ecosystems". *Biogeochemistry*, No. 1: 107–20. https://doi.org/10.1023/A:1005967203053.

Gómez, Luis, Angela Contreras, David Bolonio, Julia Quintana, Luis Oñate-Sánchez, and Irene Merino. 2019. "Chapter 10 – Phytoremediation with trees". In *Advances in Botanical Research: Regulation of Nitrogen-Fixing Symbioses in Legumes, volume 94*, edited by Pierre Fredo, Florian Frugier, and Catherine Masson-Boivin, 281–321. Cambridge, MA: Academic Press Elsevier Inc. https://doi.org/10.1016/bs.abr.2018.11.010.

Gray, E. J., and Donald L. Smith. 2005. "Intracellular and extracellular PGPR: Commonalities and distinctions in the plant–bacterium signaling processes". *Soil Biology and Biochemistry*, No. 3: 395–412. https://doi.org/10.1016/j.soilbio.2004.08.030.

Grcman, Helena, Spela Velikonja-Bolta, D. Vodnik, Be Kos, and Domen Lestan. 2001. "EDTA enhanced heavy metal phytoextraction: Metal accumulation, leaching and toxicity". *Plant Soil*, No. 235: 105–14. https://doi.org/10.1023/A:1011857303823.

Hauser, Lukas, Susan Tandy, Rainer Schulin, and Bernd Nowack. 2005. "Column extraction of heavy metals from soils using the biodegradable chelating agent EDDS". *Environmental Science & Technology*, No. 17: 6819–24. https://doi.org/10.1021/es050143r.

Heaton, A. C. P., C. L. Rugh, N.-J. Wang, and R. B. Meagher. 1998. "Phytoremediation of mercury- and methylmercury-polluted soils using genetically engineered plants". *Journal of Soil Contamination*, No. 4: 497–509. https://doi.org/10.1080/10588339891334384.

Hinsinger, Philippe, Benoît Jaillard, and Joseph E. Dufey. 1992. "Rapid weathering of a trioctahedral mica by the roots of ryegrass". *Soil Science Society of America Journal*, No. 3: 977–82. 10.2136/sssaj1992.03615995005600030049x.

Hirschi, Kendal D., Victor D. Korenkov, Nathaniel L. Wilganowski, and Goerge J. Wagner. 2000. "Expression of arabidopsis *CAX2* in tobacco: Altered metal accumulation and increased manganese tolerance". *Plant Physiology*, No. 1: 125–33. https://doi.org/10.1104/pp.124.1.125.

Huang, Jianwei W., Jianjun Chen, William R. Berti, and Scott D. Cunningham. 1997. "Phytoremediation of lead-contaminated soils: Role of synthetic chelates in lead phytoextraction". *Environmental Science & Technology*, No. 3: 800–5. https://doi.org/10.1021/es9604828.

Jensen, Lars Bogø, J. L. Ramos, Z. Kaneva, and Søeren Molin. 1993. "A substrate-dependent biological containment system for *Pseudomonas putida* based on the *Escherichia coli gef* gene". *Applied and Environmental Microbiology*, No. 11: 3713–7. https://doi.org/10.1128/AEM.59.11.3713-3717.1993.

Kaur, Ravedeep, Poonam Yadav, Sukhmeen K. Kohli, Vinod Kumar, Palak Bakshi, Bilal Ahmad Mir, Ashwani Kuma Thukral, and Renu Bhardwaj. 2018. "Chapter 4 – Emerging trends and tools in transgenic plant technology for phytoremediation of toxic metals and metalloids". In *Transgenic Plant Technology for Remediation of Toxic Metals and Metalloids*, edited by Majeti Narasimha Vara Prasad, 63–88. Cambridge, MA: Academic Press Elsevier Inc. https://doi.org/10.1016/B978-0-12-814389-6.00004-3.

Kayser, A., K. Wenger, A. Keller, W. Attinger, H. R. Felix, S. K. Gupta, and R. Schulin. 2000. "Enhancement of phytoextraction of Zn, Cd, and Cu from calcareous soil: The use of NTA and sulfur amendments". *Environmental Science & Technology*, No. 9: 1778–83. https://doi.org/10.1021/es990697s.

Kazemalilou, Z., N. Delangiz, B.A. Lajayer, and M. Ghorbanpour. 2020. "Chapter 9 – Insight into plant-bacteria-fungi interactions to improve plant performance via remediation of heavy metals: an overview". In *Molecular Aspects of Plant Beneficial Microbes in Agriculture*, edited by Vivek Sharma, Richa Salwan, and Laith Khalil Tawfeeq Al-Ani, 123–32. Cambridge, MA: Academic Press Elsevier Inc. https://doi.org/10.1016/B978-0-12-818469-1.00010-9.

Keeling, S. M., R. B. Stewart, C. W. N. Anderson, and B. H. Robinso. 2003. "Nickel and cobalt phytoextraction by the hyperaccumulator *Berkheya coddii*: Implications for polymetallic phytomining and phytoremediation". *International Journal of Phytoremediation*, No. 3: 235–44. https://doi.org/10.1080/713779223.

Komárek, Michael, Pavel Tlustoš, Jiřina Száková, and Vladislav Chrastny. 2008. "The use of poplar during a two-year induced phytoextraction of metals from contaminated agricultural soils". *Environmental Pollution*, No. 56: 27–38. https://doi.org/10.1016/j.envpol.2007.03.010.

Kong, Z., and B. R. Glick. 2017. "The role of bacteria in phytoremediation". In *Applied Bioengineering: Innovations and Future Directions*, edited by Toshiomi Yoshida, 327–53. Wiley-VCH Verlag GmbH & Co. KgaA. https://doi.org/10.1002/9783527800599.ch11.

Kopp, R. F., Lawrence B. Smart, Charles A. Maynard, J. G. Isebrands, G. A. Tuskan, and Lawrence P. Abrahamson. 2001. "The development of improved willow clones for eastern north America". *The Forestry Chronicle*, No. 2: 287–92. https://doi.org/10.5558/tfc77287-2.

Kostal, Jan, Rosanna Yang, Cindy H. Wu, Ashok Mulchandani, and Wilfrid Chen. 2004. "Enhanced arsenic accumulation in engineered bacterial cells expressing ArsR". *Applied and Environmental Microbiology*, No. 8: 4582–7. https://doi.org/10.1128/AEM.70.8.4582-4587.2004.

Lemtiri, Abdoulkacem, Amandine Liénard, Taofic Alabi, Yves Brostaux, Daniel Cluzeau, Fréderic Francis, and Gilles Colinet. 2016. "Earthworms *Eisenia fetida* affect the uptake of heavy metals by plants *Vicia faba* and *Zea mays* in metal-contaminated soils". *Applied Soil Ecology*, No. 104: 67–78. 10.1016/j.apsoil.2015.11.021.

Lombi, Enzo, Fang-Jie Zhao, S. J. Dunham, and S. P. McGrath. 2001. "Phytoremediation of heavy metal contaminated soils: Natural hyperaccumulation versus chemically enhanced phytoextraction". *Journal of Environmental Quality*, No. 6: 1919–26. https://doi.org/10.2134/jeq2001.1919.

Ludvíková, Michaela, and Miroslav Griga. 2019. "Chapter 16 – Transgenic fiber crops for phytoremediation of metals and metalloids". In *Transgenic plant technology for remediation of toxic metals and metalloids*, edited by Majeti Narasimha Vara Prasad, 341–58. Cambridge, MA: Academic Press Elsevier Inc. https://doi.org/10.1016/B978-0-12-814389-6.00016-X.

Lugtenberg, Ben, and Faina Kamilova. 2009. "Plant-growth-promoting rhizobacteria". *Annual Review of Microbiology*, No. 1: 541–56. 10.1146/annurev.micro.62.081307.162918.

Luo, Chunling, Z. G. Shen, Laiqing Lou, and Xiang-Dong Li. 2006. "EDDS and EDTA-enhanced phytoextraction of metals from artificially contaminated soil and residual effects of chelant compounds". *Environmental Pollution*, No. 144: 862–71. https://doi.org/10.1016/j.envpol.2006.02.012.

Malandrino, Mery, Agnese Giacomino, Mani Karthik, Isabella Olga Zelano, Debora Fabbri, Marco Ginepro, Roger Fuoco, Patrizia Bogani, and Ornella Abollino. 2017. "Inorganic markers profiling in wild type and genetically modified plants subjected to abiotic stresses". *Microchemical Journal*, No. 134: 87–97. 10.1016/j.microc.2017.04.023.

Mejare, Malin, and Leif Bulow. 2001. "Metal-binding proteins and peptides in bioremediation and phytoremediation of heavy metals". *Trends in Biotechnology*, No. 2: 67–73. https://doi.org/10.1016/S0167-7799(00)01534-1.

Molin, Soeren, L. B. Boe, Lars B. Jensen, C. S. Kristensen, M. Givskov, J. L. Ramos, and Asim K. Bej. 1993. "Suicidal genetic elements and their use in biological containment of bacteria". *Annual Review of Microbiology*, No. 1: 139–66. 10.1146/annurev.mi.47.100193.001035.

Molina, Lázaro, Cayo Ramos, M. Carmen Ronchel, Soeren Molin, and Juan L. Ramos. 1998. "Construction of an efficient biologically contained *Pseudomonas putida* strain and its survival in outdoor assays". *Applied and Environmental Microbiology*, No. 6: 2072–8. 10.1128/AEM.64.6.2072-2078.1998.

Newby, Deborah Trishelle, T. J. Gentry, and I. L. Pepper. 2000. "Comparison of 2,4-dichlorophenoxyacetic acid degradation and plasmid transfer in soil resulting from bioaugmentation with two different pJP4 donors". *Applied and Environmental Microbiology*, No. 8: 3399–407. https://doi.org/10.1128/AEM.66.8.3399-3407.2000.

Norvell, W. A. 1984. "Comparison of chelating agents as extractants for metals in diverse soil materials". *Soil Science Society of America Journal*, No. 6: 1285–92. https://doi.org/10.2136/sssaj1984.03615995004800060017x.

Nouri, Mohamed, and Abdelmajid Haddioui. 2016. "The remediation techniques of heavy metals contaminated soils: A review". *Maghrebian Journal of Pure and Applied Science*, No. 2: 47–58.

Peters, Robert W. 1999. "Chelant extraction of heavy metals from contaminated soils". *Journal of Hazardous Materials*, No. 1-2: 151–210. https://doi.org/10.1016/S0304-3894(99)00010-2.

Pilon-Smits, Elizabeth A. H., and Marinus Pilon. 2002. "Phytoremediation of metals using transgenic plants". *Critical Reviews in Plant Sciences*, No. 5: 439–56. https://doi.org/10.1080/0735-260291044313.

Pulford, I. D., and C. Watson. 2003. "Phytoremediation of heavy metal-contaminated land by trees – A review". *Environment International*, No. 4: 529–40. https://doi.org/10.1016/S0160-4120(02)00152-6.

Rai, Prabhat Kumar, Ki-Hyun Kim, Sang Soo Lee, and Jin-Hong Lee. 2020. "Molecular mechanisms in phytoremediation of environmental contaminants and prospects of engineered transgenic plants/microbes". *Science of The Total Environment*, No. 705: 135858. https://doi.org/10.1016/j.scitotenv.2019.135858.

Rashmi, K., Sowjanya Naga Thommandru, Maruthi P. Mohan, Vadivelu Balaji, and Govindarajulu Venkateswaran. 2004. "Bioremediation of ^{60}Co from simulated spent decontamination solutions". *Science of the Total Environment*, No. 328: 1–14. https://doi.org/10.1016/j.scitotenv.2004.02.009.

Robinson, Brett H., R. R. Brooks, and Brent E. Clothier. 1999. "Soil amendments affecting nickel and cobalt uptake by *Berkheya coddii*: Potential use for phytomining and phytoremediation". *Annals of Botany*, No. 6: 689–94. https://doi.org/10.1006/anbo.1999.0970.

Robinson, Brett H., Tessa M. Mills, Daniel Petit, Lindsay E. Fung, Steve R. Green, and Brent E. Clothier. 2000. "Natural and induced cadmium-accumulation in poplar and willow: Implications for phytoremediation". *Plant Soil*, No. 227: 301–6. https://doi.org/10.1023/A:1026515007319.

Ronchel, M. Carmen, and Juan L. Ramos. 2001. "Dual system to reinforce biological containment of recombinant bacteria designed for rhizoremediation". *Applied and Environmental Microbiology*, No. 6: 2649–56. https://doi.org/10.1128/AEM.67.6.2649-2656.2001.

Ronchel, M. Carmen, Cayo Ramos, Lars B. Jensen, Soeren Molin, and Juan L. Ramos. 1995. "Construction and behaviour of biologically contained bacteria for environmental applications in bioremediation". *Applied and Environmental Microbiology*, No. 8: 2990–4. https://doi.org/10.1128/AEM.61.8.2990-2994.1995.

Sarwar, Nadeem, Muhammad Imran, Muhammad Rashid Shaheen, Wajid Ishaque, Asif M. Kamran, Amar Matloob, Abdur Rehim, and Saddam Hussaine. 2017. "Phytoremediation strategies for soils contaminated with heavy metals: Modifications and future perspectives". *Chemosphere*, No. 171: 710–21. https://doi.org/10.1016/j.chemosphere.2016.12.116.

Singh, Shailendra, Seung Hyun Kang, Wonkyu Lee, Ashok Mulchandani, and Wilfred Wenig. 2010. "Systematic engineering of phytochelatin synthesis and arsenic transport for enhanced arsenic accumulation in *E. coli*". *Biotechnology and Bioengineering*, No. 4: 780–5. https://doi.org/10.1002/bit.22585.

Singh, Shailendra, Seung Hyun Kang, Ashok Mulchandani, and Wilfrid Chen. 2008. "Bioremediation: Environmental clean-up through pathway engineering". *Current Opinion in Biotechnology*, No. 5: 437–44. https://doi.org/10.1016/j.copbio.2008.07.012.

Strong, Lisa C., Hugh McTavish, Michael J. Sadowsky, and Lawrence P. Wackett. 2000. "Field-scale remediation of atrazine contaminated soil using recombinant *Escherichia coli* expressing atrazine chlorohydrolase". *Environmental Microbiology*, No. 1: 91–8. https://doi.org/10.1046/j.1462-2920.2000.00079.x.

Thayalakumaran, T., Brett H. Robinson, Iris Vogeler, David R. Scotter, Brent E. Clothier, and H. J. Percival. 2003. "Plant uptake and leaching of copper during EDTA-enhanced phytoremediation of repacked and undisturbed soil". *Plant Soil*, No. 2: 415–23. https://doi.org/10.1023/A:1025527931486.

Thayalakumaran, T., Iris Vogeler, David R. Scotter, H. J. Percival, Brett H. Robinson, and Brent E. Clothier. 2003a. "Leaching of copper from contaminated soil following the application of EDTA. I. Repacked soil experiments and a model". *Australian Journal of Soil Research*, No. 2: 323–33. https://doi.org/10.1071/SR02059.

Thayalakumaran, T., Iris Vogeler, David R. Scotter, H. J. Percival, Brett H. Robinson, and Brent E. Clothier. 2003b. "Leaching of copper from contaminated soil following the application of EDTA. II. Intact core experiments and model testing". *Australian Journal of Soil Research*, No. 2: 335–50. https://doi.org/10.1071/SR02060.

Torres, Begoña, Susanne Jaenecke, Kenneth N. Timmis, José Luis Garcia, and Eduardo Diaz. 2003. "A dual lethal system to enhance containment of recombinant micro-organisms". *Microbiology*, No. 149: 3595–601. https://doi.org/10.1099/mic.0.26618-0.

Vassil, Andrew D., Yoram Kapulnik, Ilya Raskin, and David E. Salt. 1998. "The role of EDTA in lead transport and accumulation by Indian mustard". *Plant Physiology*, No. 2: 447–53. https://doi.org/10.1104/pp.117.2.447.

Wei, Shuhe, Jaime A. Teixeira da Silva, and Qixing Zhou. 2008. "Agro-improving method of phytoextracting heavy metal contaminated soil". *Journal of Hazardous Materials*, No. 3: 662–8. https://doi.org/10.1016/j.jhazmat.2007.05.014.

Wu, Cindy H., Ashok Mulchandani, and Wilfred Chen. 2008. "Versatile microbial surface-display for environmental remediation and biofuel production". *Trends in Microbiology*, No. 4: 181–8. https://doi.org/10.1016/j.tim.2008.01.003.

Wu, J., F. C. Hsu, and S. D. Cunningham. 1999. "Chelate-assisted Pb phytoextraction: Pb availability, uptake, and translocation constraints". *Environmental Science & Technology*, No. 11: 1898–904. https://doi.org/10.1021/es9809253.

Zhang, Chi, Philippe Mora, Jun Dai, Xufei Chen, Syephanie Giusti-Miller, Nuria Ruiz-Camacho, Elena Velasquez, and Patrick Lavelle. 2016. "Earthworm and organic amendment effects on microbial activities and metal availability in a contaminated soil from China". *Applied Soil Ecology*, No. 104: 54–66. https://doi.org/10.1016/j.apsoil.2016.03.006.

6 Engineered and intrinsic *in situ* bioremediation of heavy metals and radionuclides

Introduction

Our knowledge of physiology, biochemistry, and genetics of bioremediation processes has reached a significant level of precision nowadays. This body of knowledge is currently oriented to apply bioremediation in the field in conjugation with process engineering. This approach exploits the innate bioactivities of indigenous microorganisms and/or plants to contain contaminants and impede their dispersion in the environment by altering their solubility or by sequestrating them. Two bioremediation techniques (*in situ* and *ex situ*) are currently applied. During *ex situ* bioremediation, the material to be treated is excavated or extracted before being treated in the vicinity of the contaminated site or far from it, while *in situ* strategy is performed directly on the spot of contamination (Francis and Nancharaiah 2015, 194). This chapter is mainly devoted to discussing the practical details of engineered and *in situ* bioremediation of some metals and radioactive contaminants. Thus, a collection of case studies performed by practitioners and researchers worldwide, tying laboratory experiences to field applications, is described. Moreover, by leaning on examples of case studies from around the world, this chapter also aims to provide the reader with a clear description of the relevance of the current applications of circular bioeconomy in the bioremediation of anthropogenic generated radionuclides and heavy metals.

In situ *bioremediation of heavy metals*

Intrinsic *in situ* bioremediation is an option that employs the innate biochemical and physiological activities of indigenous microbial species that can neutralize contaminants and counter their translocation from the source (Becker and Seagren 2010, 203).

In one *in situ* study, Groudev, Georgiev, Spasova, and Nicolova (2014, 375–8) tempted to treat a 240 m² cinnamon forest soil massively contaminated

DOI: 10.4324/9781003282600-7

with different non-ferrous metal contaminants, including Cu, Zn, and Cd, using the natural activity of the already existing microbial communities under local site conditions. It was determined that the contaminants were mainly concentrated in the superior layers of the soil called horizon A. The contaminants were drained to a deeper location of the soil designated by subhorizon B_2 in which an acidified leach solution containing ammonium, phosphate ions, and acidic organic compounds (acetate and lactate) was injected to promote the local microbial activity. Results clearly showed that the drained contaminants (heavy metals) were precipitated in subhorizon B_2 thanks to the indigenous anaerobic sulphate-reducing bacteria activity. Besides, it was shown that the obtained leach, which is rich in heterotrophic bacteria and some (*Geobacter, Shewanella*) of which have electrogenic capabilities, has successfully undergone additional treatment resulting in decreasing of the concentration of organic compounds by biodegradation in a microbial fuel cell coupled with electricity production with a maximum power density of 0.68 W/m^2.

In an *in situ* study, quantitative evaluation of engineered microbial remediation of marine sediments contaminated by oil hydrocarbons and heavy metals was conducted by Wang, He, Zou, Liu, et al. (2020, 2–6). In their study, the authors treated oil-polluted sediments from the Bohai sea (China) covering an area of 0.76 m^2 with three indigenous oil-degrading bacterial strains, namely *Acinetobacter calcium acetate, Pseudomonas putida*, and *Salfobacillus*, which were dispersed on the ocean floor in a form adsorbed on powdery/granular zeolite. In quantitative analysis of heavy metals in surface sediment samples collected after 210 days of treatment, the concentrations of Ni, Cu, Pb, Cr, and V significantly dropped by 72.6% concurrently with hydrocarbon concentration decrease as a result of bioremediation. Besides, it was interestingly found that the microbial ecology of the treated site was not affected during the period of biotreatment.

Kiikkilä, Perkiömäki, Barnette, Derome, et al. (2001, 1138–9) reported an *in situ* study of biotreatment of Cu- and Ni-contaminated sediment by mulching technique. The study was performed on the *Calluna* forest site type, covered with mulch (an equal mixture of woodchips and compost). There was a significant increase in complexed Cu concentration and a decrease in free Cu^{2+} in the soil after treatment. Moreover, it was demonstrated that the toxicity of soil to the microbial community was decreased after biotreatment.

In situ *bioremediation of radionuclides*

A great effort has been made trying to solve the problem of soil and water contamination by radionuclides, whose bioremediation efficiency is primarily tightly related to the innate capability of microbes/plants for accumulation and translocation. In this part of the chapter, some examples of successful *in situ* bioremediation studies for expurgating some radionuclides of concern from contaminated soil and water are presented. According to a study by Favas, Pratas, Varun, D'Souza, et al. (2014, 997), it was inferred that several aquatic plant species endemic to the uraniferous region of Beiras (Portugal) exhibited a significant capability of removing uranium from water through bioaccumulation. It was found that their bioremediation efficiencies were in the order *Fontinalis antipyretica* > *Callitriche stagnalis* > *Callitriche hamulate* > *Ranunculus peltatus* subsp. *Saniculifolius* > *Callitriche lusitanica* > *Ranunculus trichophyllus*. Moreover, it was demonstrated that, as a whole, submerged plants showed the best uranium incorporation, followed by rooted emergent and free-floating plants.

Groudev, Spasova, Nicolova, and Georgiev (2010, 520) succeeded in bio-stimulating the indigenous microbial community of cinnamomic forest soil contaminated by uranium and radium in uranium deposit Curilo (Western Bulgaria) by using a solution of dissolved lactate, acetate, ammonium, and phosphate. The results showed that the treatment led to removing the two contaminants through precipitation, thanks to the resident sulphate-reducing bacteria activity.

In this last part of the chapter, an example of successful *ex situ* treatment is described. In a study by Willscher, Mirgorodsky, Jablonski, Ollivier, et al. (2013, 47–52), a uranium-contaminated site located in the vicinity of Ronneburg (Germany) was treated for 100 days by a mixture of three plant species: *Helianthus annuus*, *Triticale*, and *Brassica juncea* in the presence or not of the mycorrhizal fungus *Glomus intraradices* and a mixed culture of the two bacterial strains *Streptomyces tandae* F4 and *Streptomyces acidiscabies* E13. The soil to be treated was first excavated before being planted to the three plants and then fertilized with a mixture of nitrogen, phosphor, and potassium at the rate of 100 kg/ha. Out of the three plant species tested in the study, *Triticale* yielded the highest biomass productivity (4.2 t/ha/harvest); this is mainly because of its less sensitivity towards plant competition and its innate ability to grow rapidly. By considering the translocation factor of uranium, the authors found that *Helianthus annuus* exhibited the highest uranium concentration in shoots (0.09 mg/kg) when co-cultured with the microbial mixture. Besides, in the presence of 10 kg/m^2 calcareous used as a soil amendment, *Triticale* showed the best value in uranium phytoextraction yield (260 mg/ha). It was demonstrated that soil biological and chemical amendments significantly reduced uranium concentration in seepage waters. Additionally, and interestingly, after the combustion of the harvested plants, the authors succeeded in collecting and storing between 77 and 99% of the phytoaccumulated uranium.

In another *in situ* study, the removal of nitrate from acidic nitrate-contaminated aquifer was coupled with the reduction of uranium (VI) and technetium (VII) at concentrations up to 20 μM and 0.3 nM, respectively (Istok, Senko, Krumholz, Watson, et al. 2004, 471). The authors stimulated the indigenous microbial community of the site by injecting different electron donors, including glucose, ethanol, and acetate. It was demonstrated that no uranium, nitrate, or technetium reduction was detected in the experiments without electron donor addition. However, the supplementation of organic carbon compounds significantly reduced nitrate into nitrite through denitrification concomitantly to the indirect reduction of uranium and technetium by biogenic Fe(II).

Bioremediation of radionuclides and toxic metals for sustainable circular economy

In recent years, the new circular economy concept has attracted worldwide interest as an alternative economic system to the current linear economy model of production, consumption, disposal, which has shown its limits. This concept has emerged from the idea that anthropogenic waste products can all potentially be retrieved and transformed into valuable products and/ or valorized for energy purposes by chemical and/or biological approaches (Salah-Tazdaït and Tazdaït 2020, 115–16; Saldarriaga-Hernandez, Hernandez-Vargas, Iqbal, Barceló, et al. 2020, 9). The information provided in this part of the chapter offers important insights for policy and decision-makers, industrial actors, students, and academics interested in sustainable development, renewable wastes, and environmental cleanup and management.

Kabutey, Antwi, Ding, Zhao, et al. (2019, 26831–7) have conducted a study in which a microbial fuel cell consisting of a macrophyte biocathode and a heavy metal/organic compound-polluted sediment anode was used for cleaning up heavy metals in an urban polluted river. All experiments were conducted in a fed-batch mode for 120 days in which three macrophyte species (*Limnobium laevigatum*, *Pistia stratiotes*, and *Lemna minor* L) were separately tested in three different bioreactors. The study has proven the device's efficacy in removing heavy metals from the polluted sediments, mainly through precipitation followed by recovery and bioaccumulation. The percent removal obtained for each heavy metal is as follows: 99.58% (Pb), 98.46% (Cd), 95.78% (Hg), 92.60% (Cr), 89.18% (As), and 82.28% (Zn). With regard to electricity generation, the microbial fuel cell with *Lemna minor* L gave the best voltage output (0.353 V), power density (74.16 mW/m^3), and net energy production (0.015 kWh/m^3).

An interesting example of circular zero-residue process using *Arthrospira platensis* microalgae (spirulina) as bioremediation agent was reported by Serrà, Artal, García-Amorós, Gómez, et al. (2020, 5–8). The authors have successfully combined the biotreatment of heavy metals (Cu, Ni, Zn, and Cr) contaminated wastewater with bioethanol production through fermentation and saccharification. Besides, the recovered heavy metals after biomass combustion served as Fenton-like catalysts for hazardous organic compounds degradation with an efficiency rate greater than 99%. The remaining ashes have been used as an amendment for microalgae cultivation improvement, and the cycle is complete. This circular process could be an economically feasible and cost-effective option for real-scale wastewater treatment plants.

Reference list

Becker, Jennifer G., and Eric A. Seagren. 2010. "Bioremediation of hazardous organics". In *Environmental Microbiology*, edited by Ralph Mitchell and Ji-Dong Gu, 177–212. Hoboken: Wiley-Blackwell.

Favas Paulo J. C., João Pratas, Mayank Varun, Rohan D'Souza, and Manoj S. Paul. 2014. "Accumulation of uranium by aquatic plants in field conditions: Prospects for phytoremediation". *Science of the Total Environment*, No. 470–471: 993–1002. http://doi.org/10.1016/j.scitotenv.2013.10.067.

Francis, A. J., and Y. V. Nancharaiah. 2015. "*In Situ* and *Ex Situ* bioremediation of radionuclide-contaminated soils at nuclear and NORM sites". In *Environmental Remediation and Restoration of Contaminated Nuclear and Norm Sites*, edited by Leo van Velzen, 185–236. Cambridge: Woodhead Publishing.

Groudev, Stoyan, Plamen Georgiev, Irena Spasova, and Marina Nicolova. 2014. "Decreasing the contamination and toxicity of a heavily contaminated soil by *In Situ* bioremediation". *Journal of Geochemical Exploration*, No. 144: 374–9. http://doi.org/10.1016/j.gexplo.2014.01.017.

Groudev, Stoyan, Irena Spasova, Marine Nicolova, and Plamen Georgiev. 2010. "*In Situ* bioremediation of contaminated soils in uranium deposits". *Hydrometallurgy*, No. 3–4: 518–23. https://doi.org/10.1016/j.hydromet.2010.02.027.

Istok, J. D., J. M. Senko, L. R. Krumholz, D. Watson, M. A. Bogle, A. Peacock, Y.-J. Chang, and D. C. White. 2004. "*In Situ* bioreduction of technetium and uranium in a nitrate-contaminated aquifer". *Environmental Science and Technology*, No. 2: 468–75. https://doi.org/10.1021/es034639p.

Kabutey, Felix Tetteh, Philip Antwi, Jing Ding, Qing-Liang Zhao, Frank Koblah Quashie. 2019. "Enhanced bioremediation of heavy metals and bioelectricity generation in a Macrophyte-Integrated Cathode Sediment Microbiafuel Cell (mSMFC)". *Environmental Science and Pollution Research*, No. 26: 26829–43. https://doi.org/10.1007/s11356-019-05874-9.

Kiikkilä, Oili, Jonna Perkiömäki, Matthew Barnette, John Derome, Taina Pennanen, Esa Tulisalo, and Hannu Fritze. 2001. "*In Situ* bioremediation through mulching of soil polluted by a copper-nickel smelter". *Journal of Environmental Quality*, No. 4: 1134–43. https://doi.org/10.2134/jeq2001.3041134x.

Salah-Tazdaït, Rym, and Djaber Tazdaït. 2020. "Biological systems of waste management and treatment". In *Advances in Waste-To-Energy Technologies*, edited by Rajeev Pratap Singh, Vishal Prasad, and Barkha Vaish, 115–126. Boca Raton: CRC Press.

Saldarriaga-Hernandez, Sara, Gustavo Hernandez-Vargas, Hafiz M. N. Iqbal, Damiá Barceló, and Roberto Parra-Saldívar. 2020. "Bioremediation potential of *Sargassum* sp. biomass to tackle pollution in coastal ecosystems: Circular economy approach". *Science of the Total Environment*, No. 715: 1–13. https://doi.org/10.1016/j.scitotenv.2020.136978.

Serrà, Albert, Raul Artal, Jaume García-Amorós, Elvira Gómez, and Laetitia Philippe. 2020. "Circular zero-residue process using microalgae for efficient water decontamination, biofuel production, and carbon dioxide fixation". *Chemical Engineering Journal*, No. 388: 1–10. https://doi.org/10.1016/j.cej.2020.124278.

Wang, Chuanyuan, Shijie He, Yanmei Zou, Jialin Liu, Ruxiang Zhao, Xiaonan Yin, Haijiang Zhang, and Yuanwei Li. 2020. "Quantitative evaluation of *in-situ* bioremediation of compound pollution of oil and heavy metal in sediments from the Bohai Sea, China". *Marine Pollution Bulletin*, No. 150: 1–7. https://doi.org/10.1016/j.marpolbul.2019.110787.

Willscher, S., D. Mirgorodsky, L. Jablonski, D. Ollivier, D. Merten, G. Büchel, J. Wittig, and P. Werner. 2013. "Field scale phytoremediation experiments on a heavy metal and uranium contaminated site, and further utilization of the plant residues". *Hydrometallurgy*, No. 131–132: 46–53. http://doi.org/10.1016/j.hydromet.2012.08.012.

Conclusion and perspectives

Conclusion and perspectives

Despite the formidable efforts of researchers over the years in dealing with hazardous compounds (heavy metals, radionuclides, pesticides, etc.) and their management and integrating their treatment into the circular economy, there is still much work to be done and many challenges to be overcome. In this context, governments throughout the world should agree to spend more funds on bioremediation research to achieve more advances in both the fundamental understanding of the elemental biochemical and genetic processes and also in better comprehending the actions of factors that impede the bioremediation processes both in a laboratory and *in situ/ex situ* scales. Additional efforts should also be made to enhance multidisciplinarity collaborations and international cooperation. On the other hand, special attention should be given to optimizing the bioprocesses, in terms of sustainability and effectiveness, by using modern statistical approaches (experimental designs, neuronal network designs, etc.), which are economic and time-saving. More importantly, the assessment of new, more effective and suitable microbial strains and plant species remains crucial for decontaminating the environment from organic and inorganic contaminants materials. Furthermore, improvement in process engineering will significantly help expand bioremediation application in actual field conditions.

Index

Note: Page numbers in *italics* indicates figures and page numbers in **bold** indicates tables.